技工教育"十四五"规划教材

21世纪技能创新型人才培养系列教材 　计算机系列

路由交换技术

主编／朱俊

副主编／李翠梅　黄姗姗　李文文　王来兵

参编／万梦时　史丽　葛园园

中国人民大学出版社

·北京·

图书在版编目（CIP）数据

路由交换技术 / 朱俊主编．－－北京：中国人民大学出版社，2023.7
21世纪技能创新型人才培养系列教材．计算机系列
ISBN 978-7-300-31850-9

Ⅰ.①路… Ⅱ.①朱… Ⅲ.①计算机网络－路由选择－教材②计算机网络－信息交换机－教材 Ⅳ.①TN915.05

中国国家版本馆 CIP 数据核字（2023）第 122025 号

技工教育"十四五"规划教材
21世纪技能创新型人才培养系列教材·计算机系列
路由交换技术
主　编　朱　俊
副主编　李翠梅　黄姗姗　李文文　王来兵
参　编　万梦时　史　丽　葛园园
Luyou Jiaohuan Jishu

出版发行	中国人民大学出版社
社　　址	北京中关村大街 31 号　　邮政编码　100080
电　　话	010-62511242（总编室）　　010-62511770（质管部）
	010-82501766（邮购部）　　010-62514148（门市部）
	010-62515195（发行公司）　010-62515275（盗版举报）
网　　址	http://www.crup.com.cn
经　　销	新华书店
印　　刷	北京昌联印刷有限公司
开　　本	787 mm×1092 mm　1/16　　版　次　2023 年 7 月第 1 版
印　　张	16.25　　　　　　　　　　　印　次　2024 年 7 月第 2 次印刷
字　　数	368 000　　　　　　　　　　定　价　49.00 元

版权所有　侵权必究　　印装差错　负责调换

前言

21世纪已进入计算机网络时代。随着计算机网络的极大普及，社会对网络技术工程领域高素质技术技能人才的需求日益扩大。企业信息化、社会公共服务领域信息化、电子商务、电子政务等必须以计算机网络作为支撑，网络技术工程领域高素质技术技能型人才在信息化社会中发挥着越来越重要的作用。路由交换技术是计算机网络领域中最为重要的技术之一，涉及数据包的转发、路由选择、链路负载平衡、拓扑管理等重要方面。在中国特色社会主义进入新时代的背景下，在党的二十大精神的指导下，编者深入挖掘这些关键技术的创新思想和应用价值，以满足时代发展和人们工作生活的需要。为了满足社会对计算机网络高素质技术技能人才的培养需求，实现高职高专人才培养目标，提高学生的实践和创新能力，编者在多年教学实践的基础上，结合最新教学成果，依据《国务院关于印发国家职业教育改革实施方案的通知》（国发〔2019〕4号）、《关于在院校实施"学历证书＋若干职业技能等级证书"制度试点方案》（教职成〔2019〕6号）等文件精神编写了本书。

本书按照工学结合的思路进行编写，将工作、生活中的实际工作过程与学习过程相结合，力求体现实际应用场景，把握网络技术的新面貌，讲解路由交换技术的核心概念和应用流程，重视学生的技能培养，使学生在对路由交换技术有全面的认识和了解的同时，培养学生的工作能力，即解决实际问题的能力，训练学生的数据通信组网与管理能力，提升学生的职业发展能力。本书通过介绍必要的实操步骤，可以避免因缺乏实践的指导，而造成学生只会理论不会操作；同时，大量的契合工作实际的项目教学，又可以使学生在学习的同时，体会到路由交换技术的魅力，从而加深对路由交换技术知识的理解，使学生在学习中学会工作，掌握解决实际问题的能力。

本书打破了传统的学科知识体系，以"项目引导，任务驱动"的方式编写。本书的编者有着丰富的工作实践经验，并与有丰富实践经历的企业工程技术人员密切联系、相互合作，精心设计项目、任务。本书中的项目和任务在实际应用中都能找到原型，从而为解决工作生活中的实际问题提供参考。项目设计遵循由浅入深、循序渐进的原则，每个项目贯穿实际工作过程，即项目目标→任务背景→任务规划→任务实施→任务验证→项目小结共六个步骤。因此，完成项目的过程，也就是一个完成实际工作任务的过程。

我们将实践能力的训练和理论知识的教学穿插在这六个步骤中，实践能力的训练围绕来自工作实践的工作任务来展开，技能的讲授以满足完成项目和任务为度，避免了教学过程中因过多知识的堆砌而使学生产生畏难心理，使学生能很好地将专业技能转换为工作能力，使教学内容更有实用性、针对性。学生在学习过程中，既掌握了必备的实操知识，又掌握了解决工作中实际问题的能力。同时，本书在模拟工作过程中，引导学生培养分析问题能力、组织协调能力，使学生的工作能力得到提升。

本书共 11 个项目，项目 1 介绍使用 eNSP 搭建基础网络，项目 2 介绍设备基础配置，项目 3 介绍 FTP 和 DHCP，项目 4 介绍交换技术，项目 5 介绍路由技术，项目 6 介绍网络可靠性，项目 7 介绍广域网技术，项目 8 介绍网络安全技术，项目 9 介绍 IPv6 协议，项目 10 介绍 WLAN 技术，项目 11 以某公司网络自动化运维为例，介绍网络自动化运维。本书提供了丰富的立体化教学资源，为授课教师提供数字资源、任务的操作视频，读者可以扫描书中的二维码观看操作视频。

本书着眼于路由交换技术的普及与提高，面向从事计算机网络技术相关工作的读者，既可作为高等职业学校、高等专科学校、成人高校及本科院校举办的二级职业技术学院的计算机网络技术、计算机应用技术及相关专业的教材，也可作为非计算机专业和继续教育的网络课程教材，还是一本适合广大计算机网络爱好者自学的参考书。

本书由朱俊主编，并负责全书的总体策划，完成统稿、定稿工作，李翠梅、黄姗姗、李文文、王来兵任副主编。各项目编写分工如下：项目 1、附录由朱俊、史丽共同编写，项目 2 和项目 3 由李文文、葛园园共同编写，项目 4 和项目 7 由黄姗姗、李翠梅共同编写，项目 5 和项目 9 由李翠梅、万梦时共同编写，项目 6 由葛园园、黄姗姗共同编写，项目 8 由万梦时、李文文共同编写，项目 10 和项目 11 由万梦时、史丽共同编写。本书吸纳了网络系统建设与运维职业技能等级证书（中级）的成果，教学项目设计过程中得到了多位从事本课程教学、有着丰富教学经验的一线教师的指导和帮助，在此一并表示感谢。同时，本书合作企业深圳市讯方技术股份有限公司相关人员提供了案例，并进行了审稿，在此表示感谢。此外，还要感谢中国人民大学出版社的编辑，在本书的策划与编写过程中，他们提出了很好的建议。

由于时间仓促和编者水平有限，书中难免存在不当和欠妥之处，敬请各位专家、读者批评指正，以便进一步完善。

编　者

目录

| 项目 1 | 使用 eNSP 搭建基础网络 | 1 |

  任务 搭建基础 IP 网络 ……………………………………………………… 1

| 项目 2 | 设备基础配置 | 10 |

  任务 设备基础配置 …………………………………………………………… 10

| 项目 3 | FTP 和 DHCP | 20 |

  任务 1 配置 FTP 业务 ……………………………………………………… 20
  任务 2 配置 DHCP …………………………………………………………… 27

| 项目 4 | 交换技术 | 40 |

  任务 1 交换网络基础 ………………………………………………………… 40
  任务 2 虚拟局域网技术 ……………………………………………………… 52
  任务 3 基于 STP 的可靠网络配置 ………………………………………… 68

| 项目 5 | 路由技术 | 78 |

  任务 1 静态路由和浮动静态路由 ………………………………………… 78
  任务 2 单臂路由配置 ………………………………………………………… 83
  任务 3 VLAN 间路由配置 …………………………………………………… 88
  任务 4 单区域 OSPF 配置 ……………………………………………………… 92

项目 6　网络可靠性 ·············· 99

　　任务 1　基于 VRRP 的 ISP 双出口备份链路配置 ············· 99
　　任务 2　基于 VRRP 的负载均衡出口链路配置 ············· 108
　　任务 3　链路聚合 ············· 121

项目 7　广域网技术 ·············· 131

　　任务 1　基于 PAP 认证的公司与分部安全互联 ············· 131
　　任务 2　基于 CHAP 认证的公司与分部安全互联 ············· 137
　　任务 3　基于 PPPoE 认证的公司出口配置 ············· 143

项目 8　网络安全技术 ·············· 149

　　任务 1　ACL 技术 ············· 149
　　任务 2　NAT 技术 ············· 164
　　任务 3　访问控制技术 ············· 182

项目 9　IPv6 协议 ·············· 200

　　任务 1　基于 IPv6 的静态路由 ············· 200
　　任务 2　基于 IPv6 的单臂路由 ············· 207

项目 10　WLAN 技术 ·············· 219

　　任务　基于 WLAN 的公司无线网络配置 ············· 219

项目 11　网络自动化运维 ·············· 231

　　任务　某公司网络自动化运维配置 ············· 231

参考文献 ············· 243

附录 ············· 244

　　附录 A　常用命令 ············· 244
　　附录 B　实训报告模板 ············· 251

项目 1 使用 eNSP 搭建基础网络

项目目标

1. 掌握 eNSP 模拟器的基本设置方法
2. 掌握使用 eNSP 搭建简单的端到端网络的方法
3. 掌握在 eNSP 中使用 Wireshark 捕获 IP 报文的方法

微课视频

使用 eNSP 搭建基础网络

任务 搭建基础 IP 网络

任务背景

熟悉华为 eNSP 模拟器的基本使用方法，并使用模拟器自带的抓包软件捕获网络中的报文，以便更好地理解 IP 网络的工作原理。

任务规划

通过本任务，掌握 eNSP 模拟器的基本设置方法，掌握使用 eNSP 搭建简单的端到端网络的方法，掌握在 eNSP 中使用 Wireshark 捕获 IP 报文的方法。

任务实施

步骤一 安装 eNSP

下载 eNSP 最新版本，并双击安装。
在 eNSP 中使用几个特殊设备时的操作方法如下：
（1）只要用到以下设备，都需要去官网上下载相应的镜像文件，如图 1-1 所示。
（2）例如在 eNSP 中选用 USG6000V 后，右键点击启动，如图 1-2 所示。
（3）启动设备后会弹出导入设备包的对话框，如图 1-3 所示。

软件名称	文件大小	发布时间	下载次数	下载
CE.zip	564.58MB	2019/03/08	7104	
CX.zip	405.65MB	2019/03/08	4993	
NE40E.zip	405.69MB	2019/03/08	5541	
NE5000E.zip	405.19MB	2019/03/08	4899	
NE9000.zip	405.48MB	2019/03/08	4792	
USG6000V.zip	344.93MB	2019/03/08	6885	

图 1-1 镜像文件

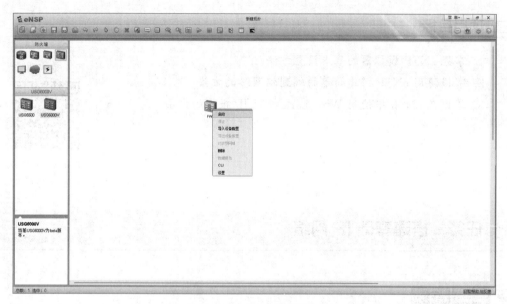

图 1-2 eNSP 主界面

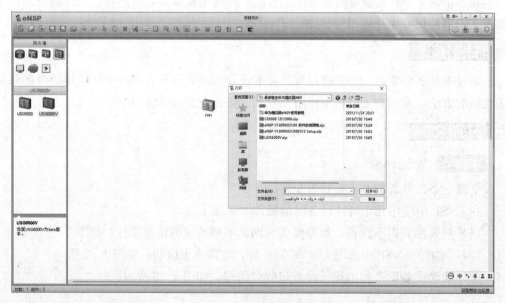

图 1-3 导入设备包的对话框

(4)单击"浏览"按钮,找到镜像文件导入即可,如图 1-4 所示。

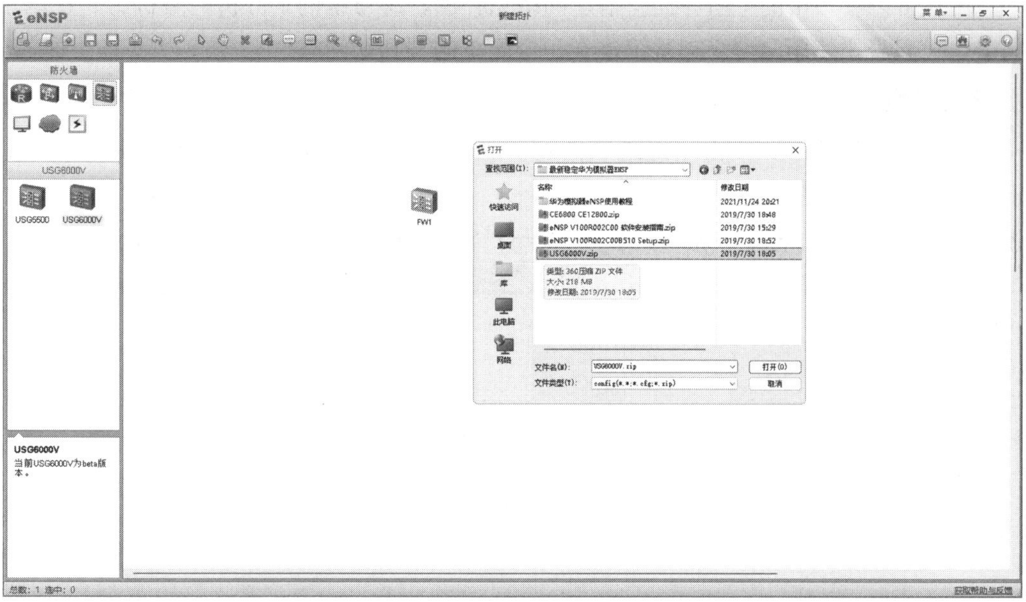

图 1-4 选择镜像文件

> **步骤二** **启动 eNSP**

本步骤介绍 eNSP 模拟器的启动与初始化界面。用户通过使用模拟器能够快速学习与掌握 TCP/IP 的原理知识,熟悉网络中的各种操作。

启动 eNSP 后,将打开启动界面,如图 1-5 所示。左侧面板中的图标代表 eNSP 所支持的各种产品及设备,中间面板则包含多种网络场景的样例。

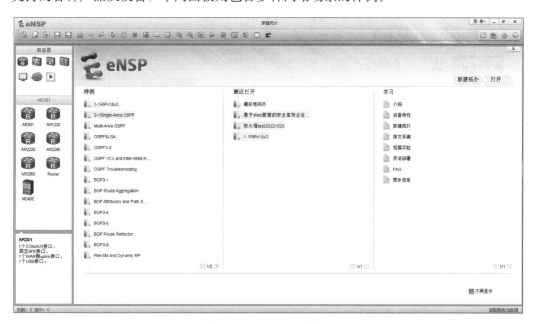

图 1-5 eNSP 启动界面

单击窗口左上角的"新建"图标,创建一个新的实验场景。

用户可以在弹出的空白界面上搭建网络拓扑图,练习组网,分析网络行为。在本示例中,我们需要使用两台终端系统建立一个简单的端到端网络。

步骤三 建立拓扑

在左侧面板顶部,单击"终端"图标。在显示的终端设备中,选中"PC"图标,把图标拖动到空白界面上,如图1-6所示。

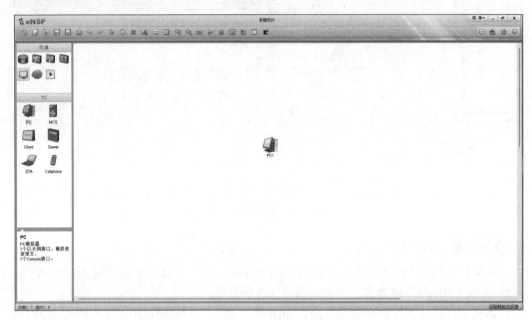

图1-6 eNSP拖动图标到空白界面

使用相同步骤,再拖动一个"PC"图标到空白界面上,建立一个端到端网络拓扑。PC设备模拟的是终端主机,可以再现真实的操作场景。

步骤四 建立一条物理连接

在左侧面板顶部,单击"设备连线"图标。在显示的媒介中,选择"Copper (Ethernet)"图标。单击图标后,光标代表一个连接器。单击客户端设备,会显示该模拟设备包含的所有端口。单击"Ethernet 0/0/1"选项,连接此端口,如图1-7所示。

单击另外一台设备并选择"Ethernet 0/0/1"端口作为该连接的终点,此时,两台设备间的连接完成。

可以观察到,在已建立的端到端网络中,连线的两端显示的是两个红点,表示该连线连接的两个端口都处于Down状态。

步骤五 进入终端系统配置界面

右击一台终端设备,在弹出的属性菜单中选择"设置"选项,查看该设备的系统配置信息,如图1-8所示。

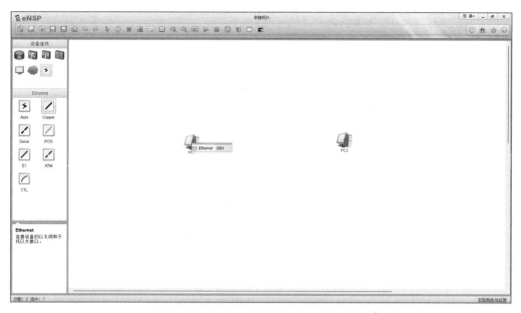

图 1-7 建立物理连接

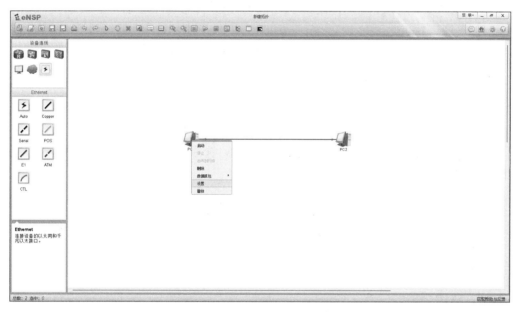

图 1-8 进入终端系统配置界面

弹出的设置属性窗口包含"基础配置""命令行""组播""UDP 发包工具"与"串口"五个标签页,分别用于不同需求的配置。

步骤六 配置终端系统

选择"基础配置"标签页,在"主机名"文本框中输入主机名称。在"IPv4 配置"区域,单击"静态"选项按钮。在"IP 地址"文本框中输入 IP 地址。建议按照图 1-9 所示配置 IP 地址及子网掩码。配置完成后,单击窗口右下角的"应用"按钮,再单击"PC1"窗口右上角的 X 关闭该窗口。

路由交换技术

图 1-9 配置终端系统

使用相同步骤配置 PC2。建议将 PC2 的 IP 地址配置为 192.168.1.2，子网掩码配置为 255.255.255.0。

完成基础配置后，两台终端系统可以成功建立端到端通信。

步骤七 启动终端系统设备

可以使用以下两种方法启动设备：

右击一台设备，在弹出的菜单中，选择"启动"选项，启动该设备。

拖动光标选中多台设备，通过右击显示菜单，选择"启动"选项，启动所有设备，如图 1-10 所示。

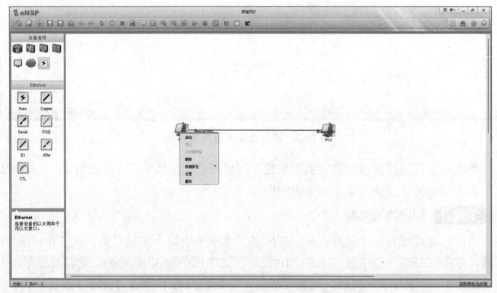

图 1-10 选中多台设备启动

设备启动后,线缆上的红点将变为绿色,表示该连接为 Up 状态。

当网络拓扑中的设备变为可操作状态后,用户可以监控物理连接中的接口状态与介质传输中的数据流。

步骤八 捕获接口报文

选中设备并右击,在显示的菜单中单击"数据抓包"选项后,会显示设备上可用于抓包的接口列表。从列表中选择需要被监控的接口,如图 1-11 所示。

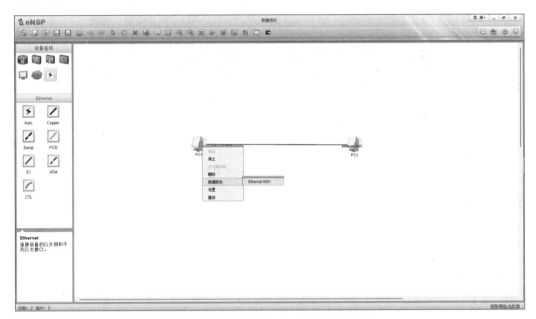

图 1-11   选择数据抓包接口

接口选择完成后,Wireshark 抓包工具会自动激活,捕获选中接口所收发的所有报文。如果需要监控更多接口,则重复上述步骤,选择不同接口即可,Wireshark 将会为每个接口激活不同实例来捕获数据包。

根据被监控设备的状态,Wireshark 可捕获选中接口上产生的所有流量,生成抓包结果。在本实例的端到端组网中,我们需要先通过配置来产生一些流量,再观察抓包结果。

步骤九 生成接口流量

我们可以使用以下两种方法打开命令行界面:

(1) 双击设备图标,在弹出的窗口中选择"命令行"标签页。

(2) 右击设备图标,在弹出的属性菜单中,选择"设置"选项,然后在弹出的窗口中选择"命令行"标签页。

产生流量最简单的方法是使用 ping 命令发送 ICMP 报文。在命令行界面输入 ping <ip address>命令,其中<ip address>设置为对端设备的 IP 地址,如图 1-12 所示。

生成的流量会在该界面的回显信息中显示,包含发送的报文和接收的报文。

生成流量之后,通过 Wireshark 捕获报文并生成抓包结果。用户可以在抓包结果中

图 1-12　使用 ping 命令发送 ICMP 报文

查看到 IP 网络的协议的工作过程，以及报文中基于 OSI 参考模型各层的详细内容。

步骤十　观察捕获的报文

查看 Wireshark 所抓取到的报文，如图 1-13 所示。

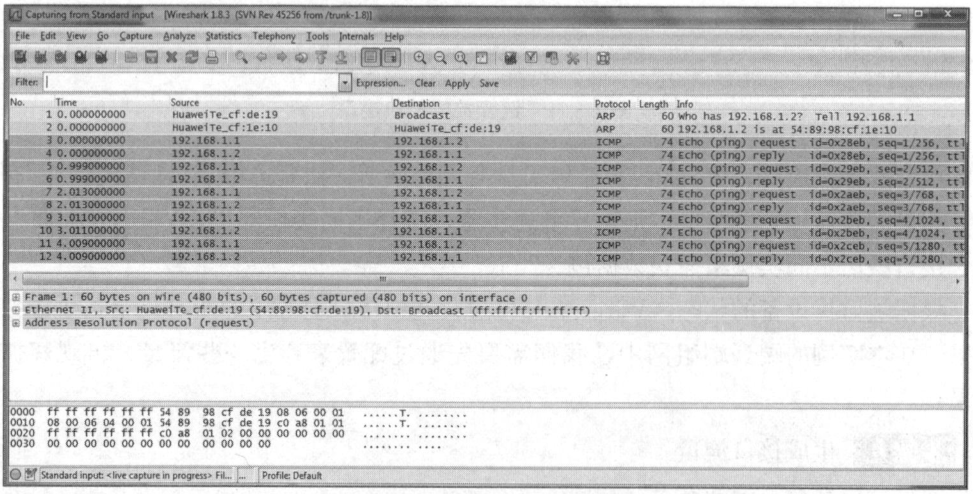

图 1-13　查看 Wireshark 所抓取到的报文

　　Wireshark 程序包含许多针对所捕获报文的管理功能。其中一个比较常用的功能是过滤功能，可用来显示某种特定报文或协议的抓包结果。用户在菜单栏下面的"Filter"文本框里输入过滤条件就可以使用该功能。最简单的过滤方法是在文本框中先输入协议名称（小写字母），再按回车键。在本示例中，Wireshark 抓取了 ICMP 与 ARP 两种协议的报文。在"Filter"文本框中输入 icmp 或 arp 再按回车键后，在回显中就将只显示 ICMP 或 ARP 报文的捕获结果。

Wireshark 界面包含三个面板，分别显示的是数据包列表、每个数据包的内容明细以及数据包对应的十六进制的数据格式。报文内容明细对于理解协议报文格式十分重要，同时也显示了基于 OSI 参考模型的各层协议的详细信息。

任务验证

无。

项目小结

通过本项目的学习，同学们了解了 eNSP 是一款界面友好、操作简单且具备极高仿真度的数据通信设备模拟器。这款仿真软件运行的是物理设备的 VRP 操作系统，最大限度地模拟真实设备环境。eNSP 可以用于模拟工程开局与网络测试，协助用户高效地构建企业优质的 ICT 网络。eNSP 支持对接真实设备和数据包的实时抓取，帮助用户深刻理解网络协议的运行原理，协助用户更好地进行网络技术的钻研和探索。另外，eNSP 贴合用户的最真实需求，用户可以利用 eNSP 模拟相关实验。

项目 2 设备基础配置

项目目标

1. 掌握设备系统参数的配置方法
2. 掌握登录的配置方法和重启设备的方法
3. 掌握配置路由器接口 IP 地址的方法
4. 掌握测试两台直连路由器连通性的方法

微课视频

设备基础配置

任务 设备基础配置

任务背景

某公司购买了两台华为 AR G3 系列路由器。在使用路由器之前,需要先配置路由器的设备名称、系统时间及登录密码等管理信息,拓扑图如图 2-1 所示。

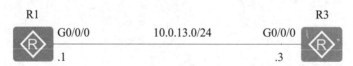

图 2-1 设备基础配置拓扑图

任务规划

通过本任务,掌握设备系统参数的配置方法,包括设备名称、系统时间及系统时区;掌握 Console 口空闲超时时长的配置方法;掌握登录信息和登录密码的配置方法;掌握保存配置文件的方法;掌握配置路由器接口 IP 地址的方法;掌握测试两台直连路由器连通性的方法;掌握重启设备的方法。

任务实施

步骤一 查看系统信息

执行 display version 命令,查看路由器的软件版本与硬件信息。

```
<Huawei>display version
Huawei Versatile Routing Platform Software
VRP (R) software, Version 5.160 (AR2200 V200R007C00SPC600)
Copyright (C) 2011-2016 HUAWEI TECH CO., LTD
Huawei AR2220E Router uptime is 0 week, 3 days, 21 hours, 43 minutes
BKP 0 version information:
......output omit......
```

命令回显信息包含 VRP 版本、设备型号和启动时间等信息。

步骤二 修改系统时间

VRP 系统会自动保存时间,但如果时间不正确,则可以在用户视图下执行 clock timezone 命令和 clock datetime 命令修改系统时间。

```
<Huawei>clock timezone Local add 08:00:00
<Huawei>clock datetime 12:00:00 2022-09-11
```

可以修改 Local 字段为当前地区的时区名称。如果当前时区位于 UTC+0 时区的西部,需要把 add 字段修改为 minus。

执行 display clock 命令查看生效的新系统时间。

```
<Huawei>display clock
2022-09-11 12:00:10
Sunday
Time Zone(Local) : UTC+ 08:00
```

步骤三 帮助功能和命令自动补全功能

在系统中输入命令时,问号是通配符,Tab 键是自动补全命令的快捷键。

```
<Huawei>display ?
  Cellular             Cellular interface
  aaa                  AAA
  access-user          User access
  accounting-scheme    Accounting scheme
  acl                  <Group>acl command group
  actived-alarm        Actived alarm
  actual               Current actual
  alarm                Alarm
  als                  Als
  antenna              Current antenna that outputting radio
  anti-attack          Specify anti-attack configurations
  ap                   <Group>ap command group
  ap-auth-mode         Display AP authentication mode
  ......output omit......
```

在输入信息后输入"?"可查看以输入字母开头的命令。例如，输入"dis?"，设备将输出所有以 dis 开头的命令。

在输入的信息后增加空格，再输入"?"，这时设备将尝试识别输入的信息所对应的命令，然后输出该命令的其他参数。例如，输入"dis?"，如果只有 display 命令是以"dis"开头的，那么设备将输出 display 命令的参数；如果以 dis 开头的命令还有其他命令，那么设备将报错。

另外，可以使用键盘上 Tab 键补全命令，比如键入"dis"后，按键盘上"Tab"键可以将命令补全为"display"。如果有多个以"dis"开头的命令存在，则在多个命令之间循环切换。

命令在不发生歧义的情况下可以使用简写，如"display"可以简写为"dis"或"disp"等，"interface"可以简写为"int"或"inter"等。

步骤四　进入系统视图

使用 system-view 命令可以进入系统视图，这样才可以配置接口、协议等内容。

```
<Huawei>system-view
Enter system view, return user view with Ctrl+Z.
```

步骤五　修改设备名称

配置设备时，为了便于区分，我们往往给设备定义不同的名称。如下我们依照实验拓扑图，修改设备名称。

修改 R1 路由器的设备名称为 R1。

```
[Huawei]sysname R1
[R1]
```

修改 R3 路由器的设备名称为 R3。

```
[Huawei]sysname R3
[R3]
```

步骤六　配置登录信息

配置登录标语信息来进行提示或进行登录警告。执行 header shell information 命令配置登录信息。

```
[R1]header shell information "Welcome to the Huawei certification lab."
```

退出路由器命令行界面，再重新登录命令行界面，查看登录信息是否已经修改。

```
[R1]quit
<R1>quit
   Configuration console exit, please press any key to log on
   Welcome to the Huawei certification lab.
<R1>
```

步骤七 配置 Console 口参数

默认情况下，通过 Console 口登录无密码，任何人都可以直接连接到设备进行配置。

为避免由此带来的风险，可以将 Console 口登录方式配置为密码认证方式，密码为密文形式的"Huawei@123"。

空闲时间是指经过没有任何操作的一定时间后，系统会自动退出该配置界面，再次登录会根据系统要求，提示输入密码进行验证。

设置空闲超时时长为 20 分钟，默认为 10 分钟。

```
[R1]user-interface console 0
[R1-ui-console0]authentication-mode password
[R1-ui-console0]set authentication password cipher
Warning: The "password" authentication mode is not secure, and it is strongly
recommended to use "aaa" authentication mode.
Enter Password(<8-128>):
Confirm password:
[R1-ui-console0] idle-timeout 20 0
```

执行 display this 命令查看配置结果。

```
[R1-ui-console0]display this
[V200R007C00SPC600]
#
user-interface con 0
 authentication-mode password
 set authentication password cipher %^%#[cR8Y%Qf_6Ra=OPEu'SFa*b$4hjW[O!/dX,6>9xW:ZQMPh6R1SbJt2SW`Y]:%^%#
 idle-timeout 20 0
user-interface vty 0
 authentication-mode aaa
 user privilege level 15
user-interface vty 1 4
#
return
```

退出系统，并使用新配置的密码登录系统。需要注意的是，在路由器第一次初始化启动时，也需要配置密码。

```
[R1-ui-console0]return
<R1>quit
    Configuration console exit, please press any key to log on Login authentication
Password:
Welcome to Huawei certification lab
<R1>
```

步骤八 配置接口 IP 地址和描述信息

配置 R1 上 GigabitEthernet 0/0/0 接口的 IP 地址。使用点分十进制格式（如 255.255.255.0）或根据子网掩码前缀长度配置子网掩码。

```
[R1]interface GigabitEthernet 0/0/0
[R1-GigabitEthernet0/0/0]ip address 10.0.13.1 24
[R1-GigabitEthernet0/0/0]description This interface connects to R3-G0/0/0
```

在当前接口视图下，执行 display this 命令查看配置结果。

```
[R1-GigabitEthernet0/0/0]display this
[V200R007C00SPC600]
#
interface GigabitEthernet0/0/0
 description This interface connects to R3-G0/0/0
 ip address 10.0.13.1 255.255.255.0
#
return
```

执行 display interface 命令查看接口信息。

```
[R1]display interface GigabitEthernet0/0/0
GigabitEthernet0/0/0 current state : UP
Line protocol current state : UP
Last line protocol up time : 2022-09-11 04:13:09
Description:This interface connects to R3-G0/0/0
Route Port,The Maximum Transmit Unit is 1500
Internet Address is 10.0.13.1/24
IP Sending Frames' Format is PKTFMT_ETHNT_2, Hardware address is 5489-9876-830b
Last physical up time   :  2022-09-11 03:24:01
Last physical down time:    2022-09-11 03:25:29
Current system time: 2022-09-11 04:15:30
Port Mode: FORCE COPPER
Speed  :   100,    Loopback: NONE
Duplex : FULL,    Negotiation: ENABLE
Mdi    : AUTO,    Clock   : -
Last 300 seconds input rate 2296 bits/sec, 1 packets/sec
Last 300 seconds output rate 88 bits/sec, 0 packets/sec
Input peak rate 7392 bits/sec,Record time: 2022-09-11 04:08:41
Output peak rate 1120 bits/sec,Record time: 2022-09-11 03:27:56
```

```
Input:   3192 packets, 895019 bytes
  Unicast:                0,      Multicast:           1592
  Broadcast:           1600,      Jumbo:                  0
  Discard:                0,      Total Error:            0
  CRC:                    0,      Giants:                 0
  Jabbers:                0,      Throttles:              0
  Runts:                  0,      Symbols:                0
  Ignoreds:               0,      Frames:                 0
Output:   181 packets, 63244 bytes
  Unicast:                0,      Multicast:              0
  Broadcast:            181,      Jumbo:                  0
  Discard:                0,      Total Error:            0
  Collisions:             0,      ExcessiveCollisions:    0
  Late Collisions:        0,      Deferreds:              0
    Input bandwidth utilization threshold : 100.00%
    Output bandwidth utilization threshold: 100.00%
    Input bandwidth utilization  : 0.01%
    Output bandwidth utilization :   0%
```

从命令回显信息中可以看到，接口的物理状态与协议状态均为 Up，表示对应的物理层与数据链路层均可用。

配置 R3 上 GigabitEthernet 0/0/0 接口的 IP 地址与描述信息。

```
[R3]interface GigabitEthernet 0/0/0
[R3-GigabitEthernet0/0/0]ip address 10.0.13.3 255.255.255.0
[R3-GigabitEthernet0/0/0]description This interface connects to R1-G0/0/0
```

配置完成后，通过执行 ping 命令测试 R1 和 R3 间的连通性。

```
<R1>ping 10.0.13.3
  PING 10.0.13.3: 56  data bytes, press CTRL_C to break
    Reply from 10.0.13.3: bytes=56 Sequence=1 ttl=255 time=35 ms
    Reply from 10.0.13.3: bytes=56 Sequence=2 ttl=255 time=32 ms
    Reply from 10.0.13.3: bytes=56 Sequence=3 ttl=255 time=32 ms
    Reply from 10.0.13.3: bytes=56 Sequence=4 ttl=255 time=32 ms
    Reply from 10.0.13.3: bytes=56 Sequence=5 ttl=255 time=32 ms
  ---10.0.13.3 ping statistics ---
    5 packet(s) transmitted
    5 packet(s) received
    0.00% packet loss
    round-trip min/avg/max = 32/32/35 ms
```

步骤九 查看当前设备上存储的文件列表

在用户视图下执行 dir 命令，查看当前目录下的文件列表。

```
<R1>dir
Directory of flash:/
  Idx  Attr  Size(Byte)     Date        Time(LMT)   FileName
   0   -rw-   1,738,816    Mar 10 2016  11:50:24    web.zip
   1   -rw-  68,288,896    Mar 10 2016  14:17:5     ar2220E-v200r007c00spc600.cc
   2   -rw-         739    Mar 10 2016  16:01:17    vrpcfg.zip
1,927,476 KB total (1,856,548 KB free)

<R3>dir
Directory of flash:/
  Idx  Attr  Size(Byte)     Date        Time(LMT)   FileName
   0   -rw-   1,738,816    Mar 10 2016  11:50:58    web.zip
   1   -rw-  68,288,896    Mar 10 2016  14:19:0     ar2220E-v200r007c00spc600.cc
   2   -rw-         739    Mar 10 2016  16:03:04    vrpcfg.zip
1,927,476 KB total (1,855,076 KB free)
```

步骤十 管理设备配置文件

执行 display saved-configuration 命令查看保存的配置文件。

```
<R1>display saved-configuration
    There is no correct configuration file in FLASH
```

系统中没有已保存的配置文件。执行 save 命令保存当前配置文件。

```
<R1>save
  The current configuration will be written to the device.
  Are you sure to continue? (y/n)[n]:y
  It will take several minutes to save configuration file, please wait............
  Configuration file had been saved successfully
  Note: The configuration file will take effect after being activated
```

重新执行 display saved-configuration 命令查看已保存的配置信息。

```
<R1>display saved-configuration
[V200R007C00SPC600]
#
 sysname R1
 header shell information "Welcome to Huawei certification lab"
#
 board add 0/1 1SA
 board add 0/2 1SA
……output omit……
```

执行 display current-configuration 命令查看当前配置信息。

```
<R1>display current-configuration
[V200R007C00SPC600]
#
 sysname R1
 header shell information "Welcome to Huawei certification lab"
#
 board add 0/1 1SA
 board add 0/2 1SA
 board add 0/3 2FE
……output omit……
```

一台路由器可以存储多个配置文件。执行 display startup 命令查看下次启动时使用的配置文件。

```
<R3>display startup
MainBoard:
  Startup system software:                    flash:/AR2220E-v200R007C00SPC600.cc
  Next startup system software:               flash:/AR2220E-V200R007C00SPC600.cc
  Backup system software for next startup:    null
  Startup saved-configuration file:           null
  Next startup saved-configuration file:      flash:/vrpcfg.zip
  Startup license file:                       null
  Next startup license file:                  null
  Startup patch package:                      null
  Next startup patch package:                 null
  Startup voice-files:                        null
  Next startup voice-files:                   null
```

删除闪存中的配置文件。

```
<R1>reset saved-configuration
This will delete the configuration in the flash memory.
The device configurations will be erased to reconfigure.
Are you sure? (y/n)[n]:y
  Clear the configuration in the device successfully.

<R3>reset saved-configuration
This will delete the configuration in the flash memory.
The device configurations will be erased to reconfigure.
Are you sure? (y/n)[n]:y
  Clear the configuration in the device successfully.
```

步骤十一　重启设备

执行 reboot 命令重启路由器。

```
<R1>reboot
Info: The system is now comparing the configuration, please wait.
Warning: All the configuration will be saved to the next startup configuration.
Continue ? [y/n]:n
System will reboot! Continue ? [y/n]:y
Info: system is rebooting ,please wait...
```

```
<R3>reboot
Info: The system is now comparing the configuration, please wait.
Warning: All the configuration will be saved to the next startup configuration.
Continue ? [y/n]:n
System will reboot! Continue ? [y/n]:y
```

系统提示是否保存当前配置，可根据实验要求决定是否保存当前配置。如果无法确定是否保存，则不保存当前配置。

任务验证

查看配置文件：

```
[R1]display current-configuration
[V200R007C00SPC600]
#
 sysname R1
 header shell information "Welcome to Huawei certification lab"
#
interface GigabitEthernet0/0/0
 description This interface connects to R3-G0/0/0
 ip address 10.0.13.1 255.255.255.0
#
user-interface con 0
 authentication-mode password
 set authentication password cipher %$%$4D0K*-E"t/I7[{HD~kgW,%dgkQQ! &|;XT-Dq9SFQJ.27M%dj,%$%$
 idle-timeout 20 0
#
return
```

```
[R3]dispay current-configuration
[V200R007C00SPC600]
#
 sysname R3
#
interface GigabitEthernet0/0/0
 description This interface connect to R1-G0/0/0
 ip address 10.0.13.3 255.255.255.0
#
user-interface con 0
 authentication-mode password
 set authentication password cipher %$%$M8\HO3:72:ERQ8JLoHU8,%t+lE:$9=a7"8%yMoARB]$B%t.,%$%$
 user-interface vty 0 4
#
return
```

项目小结

通过本项目的学习，同学们了解了设备的基础配置，包括设备名称、系统时间及系统时区的配置，Console 口空闲超时时长的配置，登录信息和登录密码的配置，保存配置文件的方法以及路由器接口 IP 地址的配置；学会了使用 ping 命令测试两台直连路由器的连通性以及重启设备的方法。设备使用之前需要先进行基础配置，掌握了这些基础配置方便后期设备的使用和维护。

项目 3　FTP 和 DHCP

项目目标

1. 理解建立 FTP 连接的过程
2. 掌握 FTP 服务器参数的配置
3. 掌握与 FTP 服务器传输文件的方法
4. 掌握 DHCP 全局地址池的配置方法
5. 掌握 DHCP 接口地址池的配置方法
6. 掌握在交换机端口启用 DHCP 发现功能和 IP 地址分配功能的方法

任务 1　配置 FTP 业务

任务背景

现在需要在某公司网络上配置 FTP 业务。需要把一台路由器配置为 FTP 服务器，客户端可以通过 TCP 连接与 FTP 服务器之间传输文件，拓扑图如图 3-1 所示。

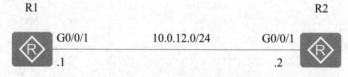

图 3-1　配置 FTP 业务实验拓扑图

任务规划

通过本任务的学习，掌握建立 FTP 连接的过程、FTP 服务器参数的配置以及与 FTP 服务器传输文件的方法。

项目 3 FTP 和 DHCP

任务实施

步骤一 实验环境准备

如果本任务使用的是空配置设备,那么请从步骤一开始配置。如果使用的设备包含上一个实验的配置,请直接从步骤二开始配置。

```
<Huawei>system-view
Enter system view, return user view with Ctrl+Z.
[Huawei]sysname R1
[R1]interface  GigabitEthernet 0/0/1
[R1-GigabitEthernet0/0/1]ip address 10.0.12.1 24
```

```
<Huawei>system-view
Enter system view, return user view with Ctrl+Z.
[Huawei]sysname R2
[R2]interface  GigabitEthernet 0/0/1
[R2-GigabitEthernet0/0/1]ip address 10.0.12.2 24
```

测试 R1 和 R2 之间的连通性。

```
[R1]ping 10.0.12.2
PING 10.0.12.2: 56   data bytes, press CTRL_C to break
  Reply from 10.0.12.2: bytes=56 Sequence=1 ttl=255 time=10 ms
  Reply from 10.0.12.2: bytes=56 Sequence=2 ttl=255 time=1 ms
  Reply from 10.0.12.2: bytes=56 Sequence=3 ttl=255 time=1 ms
  Reply from 10.0.12.2: bytes=56 Sequence=4 ttl=255 time=10 ms
  Reply from 10.0.12.2: bytes=56 Sequence=5 ttl=255 time=1 ms
---10.0.12.2 ping statistics ---
  5 packet(s) transmitted
  5 packet(s) received
  0.00%  packet loss
round-trip min/avg/max =1/4/10 ms
```

步骤二 在路由器上启用 FTP 业务

默认情况下,路由器的 FTP 功能并未启用。使用 FTP 业务之前,必须先启用 FTP 功能。配置 R1 为 FTP 服务器,R2 为客户端。

```
[R1]ftp server enable
Info: Succeeded in starting the FTP server
[R1]set default ftp-directory flash:/
```

通过在 AAA 中设置用户名和密码,授权 FTP 合法用户连接到 FTP 服务器。这样,非法用户就无法连接 FTP 服务器,降低了安全风险。

```
[R1]aaa
[R1-aaa]local-user huawei password    cipher    huawei123
Info: Add a new user.
[R1-aaa]local-user huawei service-type ftp
Info: The cipher password has been changed to an irreversible-cipher password.
Warning: The user access modes include Telnet, FTP or HTTP, and so security
risks exist.
Info: After you change the rights (including the password, access type, FTP di-
rectory, and level) of a local user, the rights of users already online do not
change. The change takes effect to users who go online after the change.
[R1-aaa]local-user huawei privilege level 15
Info: After you change the rights (including the password, access type, FTP di-
rectory, and level) of a local user, the rights of users already online do not
change. The change takes effect to users who go online after the change.
[R1-aaa]local-user huawei ftp-directory flash:
Info: After you change the rights (including the password, access type, FTP di-
rectory, and level) of a local user, the rights of users already online do not
change. The change takes effect to users who go online after the change.

[R1]display ftp-server
  FTP server is running
  Max user number                 5
  User count                      0
  Timeout value(in minute)        30
  Listening port                  21
  Acl number                      0
  FTP server's source address     0.0.0.0
```

配置完成后，可以看到 R1 为 FTP 服务器，默认情况下监听 TCP 21 号端口。

步骤三 建立 FTP 客户端与服务器的连接

建立从客户端（R2）到 FTP 服务器（R1）的连接。

```
<R2>ftp 10.0.12.1
Trying 10.0.12.1 ...
Press CTRL+K to abort
Connected to 10.0.12.1.
220 FTP service ready.
User(10.0.12.1:(none)):huawei
331 Password required for huawei.
Enter password:
230 User logged in.
[R2-ftp]
```

输入正确的用户名和密码后，可以成功登录 FTP 服务器。

下载文件前或者上传文件后，执行 dir 命令查看文件的详细信息。

```
[R2-ftp]dir
200 Port command okay.
150 Opening ASCII mode data connection for * .
drwxrwxrwx 1 noone nogroup           0 May 03 18:03 .
-rwxrwxrwx 1 noone nogroup   114552448 Jan 19 2012 AR2220E-V200R006C10SPC300.cc
-rwxrwxrwx 1 noone nogroup      159858 May 03 17:59 mon_file.txt
-rwxrwxrwx 1 noone nogroup      304700 Mar 03 11:11 sacrule.dat
-rwxrwxrwx 1 noone nogroup         783 Mar 03 11:12 default_local.cer
-rwxrwxrwx 1 noone nogroup           0 Dec 20  2015 brdxpon_snmp_cfg.efs
-rwxrwxrwx 1 noone nogroup         777 May 03 18:03 vrpcfg.zip
drwxrwxrwx 1 noone nogroup           0 Mar 10 11:14 update
drwxrwxrwx 1 noone nogroup           0 May 03 18:03 localuser
drwxrwxrwx 1 noone nogroup           0 Mar 17 10:45 dhcp
-rwxrwxrwx 1 noone nogroup         460 May 03 18:03 private-data.txt
-rwxrwxrwx 1 noone nogroup   126352896 Mar 10 11:09 AR2220E-V200R007C00SPC600.cc
drwxrwxrwx 1 noone nogroup           0 Mar 10 11:15 shelldir
-rwxrwxrwx 1 noone nogroup       11606 May 03 18:00 mon_lpu_file.txt
drwxrwxrwx 1 noone nogroup           0 Mar 18 14:45 huawei
-rwxrwxrwx 1 noone nogroup         120 Mar 18 15:02 text.txt
226 Transfer complete.
FTP: 1112 byte(s) received in 0.134 second(s) 8.29Kbyte(s)/sec.
```

配置文件的传输模式。

```
[R2-ftp]binary
200 Type set to I.
```

在 FTP 服务器上下载文件。

```
[R2-ftp]get vrpcfg.zip vrpnew.zip
200 Port command okay.
150 Opening BINARY mode data connection for vrpcfg.zip.
226 Transfer complete.
FTP: 120 byte(s) received in 0.678 second(s) 176.99byte(s)/sec.
```

从 FTP 服务器上下载文件后，执行 bye 命令关闭连接。

```
[R2-ftp]bye
221 Server closing.

<R2>dir
```

```
Directory of flash:/
  Idx  Attr   Size(Byte)  Date          Time(LMT)   FileName
  0    -rw-   114,552,448 Jan 19 2012   15:32:52    AR2220E-V200R006C10SPC300.cc
  1    -rw-       270,176 Apr 30 2016   03:17:08    mon_file.txt
  2    -rw-       304,700 Mar 03 2016   11:11:44    sacrule.dat
  3    -rw-           783 Mar 03 2016   11:12:22    default_local.cer
  4    -rw-             0 Dec 20 2015   00:06:14    brdxpon_snmp_cfg.efs
  5    -rw-           775 Apr 29 2016   17:51:48    vrpcfg.zip
  6    drw-             - Mar 10 2016   11:28:46    update
  7    drw-             - Apr 23 2016   17:33:38    localuser
  8    drw-             - Mar 21 2016   20:59:46    dhcp
  9    -rw-           394 Apr 29 2016   17:51:50    private-data.txt
  10   -rw-   126,352,896 Mar 10 2016   11:14:40    AR2220E-V200R007C00SPC600.cc
  11   drw-             - Mar 10 2016   11:29:20    shelldir
  12   -rw-        23,950 Apr 27 2016   16:06:06    mon_lpu_file.txt
  13   -rw-           120 Mar 24 2016   11:45:44    huawei.zip
  14   -rw-           777 May 10 2016   14:23:43    vrpnew.zip
```

可以通过 put 命令把一个文件上传到 FTP 服务器，上传的同时也可以为该文件配置新的文件名。

```
[R2-ftp]put vrpnew.zip vrpnew2.zip
200 Port command okay.
150 Opening BINARY mode data connection for vrpnew2.zip.
226 Transfer complete.
FTP: 120 byte(s) sent in 0.443 second(s) 270.88byte(s)/sec.
```

上传文件后，执行 dir 命令查看文件是否存在于 FTP 服务器上。

```
<R1>dir
Directory of flash:/
  Idx  Attr   Size(Byte)  Date          Time(LMT)   FileName
  0    -rw-       286,620 Mar 14 2016   09:22:20    sacrule.dat
  1    -rw-       512,000 Mar 28 2016   14:39:16    mon_file.txt
  2    -rw-     1,738,816 Mar 17 2016   12:05:36    web.zip
  3    -rw-        48,128 Mar 10 2016   14:16:56    ar2220E_v200r001sph001.pat
  4    -rw-           120 Mar 28 2016   10:09:50    iascfg.zip
  5    -rw-           699 Mar 28 2016   17:52:38    vrpcfg.zip
  6    -rw-    93,871,872 Mar 14 2016   09:13:26    ar2220E-V200R007C00SPC600.cc
  7    -rw-       512,000 Mar 28 2016   14:40:20    mon_lpu_file.txt
  8    -rw-           699 Mar 02 2016   15:44:16    vrpnew2.zip
```

分别在 R1 和 R2 上删除创建的 vrpnew2.zip 和 vrpnew.zip 文件。

```
<R1>delete flash:/vrpnew2.zip
Delete flash:/vrpnew2.zip? (y/n)[n]:y
Info: Deleting file flash:/vrpnew2.zip...succeed.

<R2>delete flash:/vrpnew.zip
Delete flash:/vrpnew.zip? (y/n)[n]:y
Info: Deleting file flash:/vrpnew.zip...succeed.
```

注意：删除配置文件时，请慎重执行，避免删除 R1 和 R2 上的整个 flash:/目录。

任务验证

查看配置文件：

```
<R1>display current-configuration
[V200R007C00SPC600]
#
 sysname R1
 ftp server enable
 set default ftp-directory flash:
#
aaa
 authentication-scheme default
 authorization-scheme default
 accounting-scheme default
 domain default
 domain default_admin
 local-user admin password cipher %$%$=i~>Xp&aY+*2cEVcS-A23Uwe%$%$
 local-user admin service-type http
 local-user huawei password cipher %$%$f+~&ZkCn]NUX7m.t;tF9R48s%$%$
 local-user huawei privilege level 15
 local-user huawei ftp-directory flash:/
 local-user huawei service-type ftp
#
interface GigabitEthernet0/0/1
 ip address 10.0.12.1 255.255.255.0
#
user-interface con 0
 authentication-mode password
```

```
   set authentication password cipher %$%$+L'YR&IZt'4,)>-*#1H",}%K-oJ_M9+'
1OU~bD (\WTqB}%N,%$%$
 user-interface vty 0 4
 #
 return

<R2>display current-configuration
[V200R007C00SPC600]
#
 sysname R2
 ftp server enable
 set default ftp-directory flash:
#
aaa
 authentication-scheme default
 authorization-scheme default
 accounting-scheme default
 domain default
 domain default_admin
 local-user admin password cipher %$%$=i~>Xp&aY+*2cEVcS-A23Uwe%$%$
 local-user admin service-type http
 local-user huawei password cipher %$%$<;qM3D/O;ZLqy/"&6wEESdg$%$%$
 local-user huawei privilege level 15
 local-user huawei ftp-directory flash:/
 local-user huawei service-type ftp
#
interface GigabitEthernet0/0/1
 ip address 10.0.12.2 255.255.255.0
#
user-interface con 0
 authentication-mode password
 set authentication password cipher %$%$1=cd%b%/O%Id-8X:by1N,+s}'4wD6TvO<
I|/pd##44C@+s#,%$%$
 user-interface vty 0 4
 #
 return
```

任务 2　配置 DHCP

配置 DHCP

任务背景

某公司网络需要配置 DHCP 业务,将网关路由器 R1 和 R3 配置为 DHCP 服务器,并配置全局地址池和接口地址池,为接入层设备分配 IP 地址,拓扑图如图 3-2 所示。

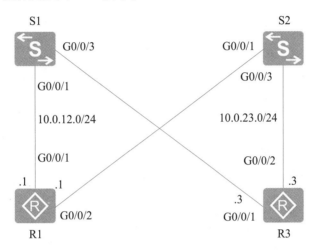

图 3-2　配置 DHCP 实验拓扑图

任务规划

通过本任务,掌握 DHCP 全局地址池的配置方法、DHCP 接口地址池的配置方法、在交换机端口启用 DHCP 发现功能和 IP 地址分配功能的方法。

任务实施

步骤一　任务环境准备

如果本任务使用的是空配置设备,需要从步骤一开始,并跳过步骤二。如果使用的设备包含上一个实验的配置,请直接从步骤二开始。

按照实验拓扑图进行基础配置以及 IP 编址,暂时关闭 R1 上的 G0/0/2 接口和 R3 上的 G0/0/1 接口。

```
<Huawei>system-view
Enter system view, return user view with Ctrl+Z.
[Huawei]sysname R1
[R1]interface GigabitEthernet 0/0/1
```

```
[R1-GigabitEthernet0/0/1]ip address 10.0.12.1 24
[R1-GigabitEthernet0/0/1]quit
[R1]interface GigabitEthernet 0/0/2
[R1-GigabitEthernet0/0/2]ip address 10.0.23.1 24
[R1-GigabitEthernet0/0/2]shutdown
[R1-GigabitEthernet0/0/2]quit
```

```
<Huawei>system-view
Enter system view, return user view with Ctrl+Z.
[Huawei]sysname R3
[R3]interface GigabitEthernet 0/0/1
[R3-GigabitEthernet0/0/1]ip address 10.0.12.3 24
[R3-GigabitEthernet0/0/1]shutdown
[R3-GigabitEthernet0/0/1]quit
[R3]interface GigabitEthernet 0/0/2
[R3-GigabitEthernet0/0/2]ip address 10.0.23.3 24
```

```
<Quidway>system-view
Enter system view, return user view with Ctrl+Z.
[Quidway]sysname S1
```

```
<Quidway>system-view
Enter system view, return user view with Ctrl+Z.
[Quidway]sysname S2
```

步骤二 清除设备上已有的配置

重新开启 R3 上的 G0/0/2 接口。

```
[R3]interface GigabitEthernet 0/0/2
[R3-GigabitEthernet0/0/2]undo shutdown
```

步骤三 进行其他准备配置

关闭 S1 和 S2 上其他无关接口。

```
[S1]interface GigabitEthernet 0/0/9
[S1-GigabitEthernet0/0/9]shutdown
[S1-GigabitEthernet0/0/9]quit
[S1]interface GigabitEthernet 0/0/10
[S1-GigabitEthernet0/0/10]shutdown
[S1-GigabitEthernet0/0/10]quit
[S1]interface GigabitEthernet 0/0/13
[S1-GigabitEthernet0/0/13]shutdown
```

```
[S1-GigabitEthernet0/0/13]quit
[S1]interface GigabitEthernet 0/0/14
[S1-GigabitEthernet0/0/14]shutdown

[S2]interface GigabitEthernet 0/0/6
[S2-GigabitEthernet0/0/6]shutdown
[S2]interface GigabitEthernet 0/0/7
[S2-GigabitEthernet0/0/7]shutdown
[S2-GigabitEthernet0/0/7]quit
[S2]interface GigabitEthernet 0/0/9
[S2-GigabitEthernet0/0/9]shutdown
[S2-GigabitEthernet0/0/9]quit
[S2]interface GigabitEthernet 0/0/10
[S2-GigabitEthernet0/0/10]shutdown
[S2-GigabitEthernet0/0/10]quit

[R1]interface GigabitEthernet 0/0/2
[R1-GigabitEthernet0/0/2]ip address 10.0.23.1 24
[R1-GigabitEthernet0/0/2]shutdown
```

确认 S1 上的 G0/0/9、G0/0/10、G0/0/13、G0/0/14 接口已关闭,S2 上的 G0/0/6、G0/0/7、G0/0/9、G0/0/10 接口已关闭。

```
<S1>display interface brief
…output omit…
Interface              PHY     Protocol   InUti   OutUti   inErrors   outErrors
GigabitEthernet0/0/1   up      up         0.01%   0.01%    0          0
GigabitEthernet0/0/2   up      up         0.01%   0.01%    0          0
GigabitEthernet0/0/3   down    down       0%      0%       0          0
GigabitEthernet0/0/4   up      up         0%      0.01%    0          0
GigabitEthernet0/0/5   up      up         0%      0.01%    0          0
GigabitEthernet0/0/6   down    down       0%      0%       0          0
GigabitEthernet0/0/7   down    down       0%      0%       0          0
GigabitEthernet0/0/8   down    down       0%      0%       0          0
GigabitEthernet0/0/9   *down   down       0%      0%       0          0
GigabitEthernet0/0/10  *down   down       0%      0%       0          0
GigabitEthernet0/0/11  down    down       0%      0%       0          0
GigabitEthernet0/0/12  down    down       0%      0%       0          0
GigabitEthernet0/0/13  *down   down       0%      0%       0          0
GigabitEthernet0/0/14  *down   down       0%      0%       0          0
…output omit…
```

```
<S2>display interface brief
…output omit…
GigabitEthernet0/0/1      up      up       0%      4.06%    0    0
GigabitEthernet0/0/2      up      up       0%      4.06%    0    0
GigabitEthernet0/0/3      up      up       0%      4.06%    0    0
GigabitEthernet0/0/4      up      up       0%      20.40%   0    0
GigabitEthernet0/0/5      up      up       0%      20.40%   0    0
GigabitEthernet0/0/6      *down   down     0%      2.04%    0    0
GigabitEthernet0/0/7      *down   down     2.03%   2.03%    0    0
GigabitEthernet0/0/8      down    down     0%      0%       0    0
GigabitEthernet0/0/9      *down   down     1.91%   1.91%    0    0
GigabitEthernet0/0/10     *down   down     3.95%   0.12%    0    0
GigabitEthernet0/0/11     up      up       0%      4.06%    0    0
GigabitEthernet0/0/12     up      up       0%      4.06%    0    0
…output omit…
```

确认 R1 上只有 G0/0/2 接口被关闭，R3 上只有 G0/0/1 接口被关闭。

```
<R1>display ip interface brief
…output omit…
GigabitEthernet0/0/1      10.0.12.1/24     up       up
GigabitEthernet0/0/2      10.0.23.1/24     *down    down
…output omit…
```

```
<R3>display ip interface brief
…output omit…
GigabitEthernet0/0/1      10.0.12.3/24     *down    down
GigabitEthernet0/0/2      10.0.23.3/24     up       up
…output omit…
```

步骤四 启用 DHCP 功能

默认情况下，DHCP 功能并未启用。在路由器上启用 DHCP 功能。

```
[R1]dhcp enable
```

```
[R3]dhcp enable
```

步骤五 创建全局 IP 地址池

在 R1 和 R3 上分别创建名为 pool1 和 pool2 的地址池，并配置地址池中地址的起始范围、网关地址和地址租期。

```
[R1]ip pool pool1
Info: It's successful to create an IP address pool.
[R1-ip-pool-pool1]network 10.0.12.0 mask 24
[R1-ip-pool-pool1]gateway-list 10.0.12.1
[R1-ip-pool-pool1]lease day 1 hour 12
[R1]interface GigabitEthernet 0/0/1
[R1-GigabitEthernet0/0/1]dhcp select global

[R3]ip pool pool2
Info: It's successful to create an IP address pool.
[R3-ip-pool-pool2]network 10.0.23.0 mask 24
[R3-ip-pool-pool2]gateway-list 10.0.23.3
[R3-ip-pool-pool2]lease day 1 hour 12
[R3]interface GigabitEthernet 0/0/2
[R3-GigabitEthernet0/0/2]dhcp select global
```

在路由器上执行 display ip pool name＜name＞命令，查看配置的 IP 地址池中的参数。

```
<R1>display ip pool name pool1
  Pool-name       : pool1
  Pool-No         : 0
  Lease           : 1 Days 12 Hours 0 Minutes
  Domain-name     : -
  DNS-server0     : -
  NBNS- server0:  -
  Netbios- type:  -
  Position        : Local          Status          : Unlocked
  Gateway- 0      : 10.0.12.1
  Network         : 10.0.12.0
  Mask            : 255.255.255.0
  VPN instance    : --
  ---------------------------------------------------------------
  Start         End         Total  Used  Idle(Expired)  Conflict  Disable
  ---------------------------------------------------------------
  10.0.12.1   10.0.12.254   253    0     253(0)         0         0
  ---------------------------------------------------------------
```

配置 S1 通过缺省管理端口 VLANIF 1 向 DHCP 服务器（R1）申请 IP 地址。在 S2 上使用相同配置向 R3 申请 IP 地址。

```
[S1]dhcp enable
[S1]interface Vlanif 1
[S1-Vlanif1]ip address dhcp-alloc

<S1>display ip interface brief
···output omit···
Interface      IP Address/Mask    Physical    Protocol
MEth0/0/1      unassigned         down        down
NULL0          unassigned         up          up(s)
Vlanif1        10.0.12.254/24     up          up
```

验证 S1 从 R1 上名为 pool1 的 DHCP 地址池获取 IP 地址,S2 从 R3 上名为 pool2 的 DHCP 地址池获取 IP 地址。

```
<R1>display ip pool name pool1
  Pool-name      : pool1
  Pool-No        : 0
  Lease          : 1 Days 12 Hours 0 Minutes
  Domain-name    : -
  DNS-server0    : -
  NBNS-server0   : -
  Netbios-type   : -
  Position       : Local         Status        : Unlocked
  Gateway-0      : 10.0.12.1
  Network        : 10.0.12.0
  Mask           : 255.255.255.0
  VPN instance   : --
  ----------------------------------------------------------------
  Start       End          Total   Used   Idle(Expired)   Conflict   Disable
  ----------------------------------------------------------------
  10.0.12.1   10.0.12.254  253     1      252(0)          0          0
  ----------------------------------------------------------------

<R3>display ip pool name pool2
  Pool-name      : pool2
  Pool-No        : 0
  Lease          : 1 Days 12 Hours 0 Minutes
  Domain-name    : -
  DNS-server0    : -
```

```
  NBNS-server0: -
  Netbios-type: -
  Position       : Local
  Status         : Unlocked
  Gateway-0      : 10.0.23.3
  Network        : 10.0.23.0
  Mask           : 255.255.255.0
  VPN instance   : --
 ---------------------------------------------------------------------
  Start        End          Total  Used   Idle(Expired)  Conflict  Disable
 ---------------------------------------------------------------------
  10.0.23.1    10.0.23.254  253    1      252(0)         0         0
 ---------------------------------------------------------------------
```

进行新的配置前，确保 R1 和 R3 上的全局地址池配置已经完成。

步骤六　创建接口地址池

关闭 R1 上的 G0/0/1 接口、R3 上的 G0/0/2 接口。

```
[R1]interface GigabitEthernet 0/0/1
[R1-GigabitEthernet0/0/1]shutdown
```

```
[R3]interface GigabitEthernet 0/0/2
[R3-GigabitEthernet0/0/2]shutdown
```

执行 dhcp select interface 命令开启接口的 DHCP 服务功能，指定路由器从接口地址池分配地址。此时，我们还不希望激活网络中的 DHCP 服务，所以先不用开启这两个接口。

```
[R1]interface GigabitEthernet 0/0/2
[R1-GigabitEthernet0/0/2]dhcp select interface
```

```
[R3]interface GigabitEthernet 0/0/1
[R3-GigabitEthernet0/0/1]dhcp select interface
```

从 R1 和 R3 的接口地址池中为 DNS 业务预留 IP 地址，并设置接口地址池的地址租期。

```
[R1-GigabitEthernet0/0/2]dhcp server dns-list 10.0.23.254
[R1-GigabitEthernet0/0/2]dhcp server excluded-ip-address 10.0.23.254
[R1-GigabitEthernet0/0/2]dhcp server lease day 1 hour 12
```

```
[R3-GigabitEthernet0/0/1]dhcp server dns-list 10.0.12.254
[R3-GigabitEthernet0/0/1]dhcp server excluded-ip-address 10.0.12.254
[R3-GigabitEthernet0/0/1]dhcp server lease day 1 hour 12
```

在路由器上执行 display ip pool interface 命令，查看配置的接口地址池参数。此处以 R1 为例。

```
<R1>display ip pool interface GigabitEthernet 0/0/2
  Pool-name      : GigabitEthernet0/0/2
  Pool-No        : 1
  Lease          : 1 Days 12 Hours 0 Minutes
  Domain-name    : -
  DNS-server0    : 10.0.23.254
  NBNS-server0   : -
  Netbios-type   : -
  Position       : Interface         Status          : Unlocked
  Gateway-0      : 10.0.23.1
  Network        : 10.0.23.0
  Mask           : 255.255.255.0
  VPN instance   : --
  ---------------------------------------------------------------
  Start       End         Total   Used  Idle(Expired)   Conflict  Disable
  ---------------------------------------------------------------
  10.0.23.1   10.0.23.254  253    0     252(0)          0         1
  ---------------------------------------------------------------
```

关闭 S2 上 VLANIF 1 接口以清除接口现有的 IP 地址，然后重新开启此接口以便重新从 R1 的接口地址池中获取新的 IP 地址。

```
[S2]interface Vlanif 1
[S2-Vlanif1]shutdown
[S2-Vlanif1]undo shutdown
```

开启 R1 的 G0/0/2 接口，使 R1 可以通过此接口从接口地址池中分配 IP 地址。

```
[R1]interface GigabitEthernet0/0/2
[R1-GigabitEthernet0/0/2]undo shutdown
```

验证 R1 从接口地址池中为 S2 的 VLANIF 1 接口分配了新的 IP 地址。

```
<R1>display ip pool interface GigabitEthernet 0/0/2
  Pool-name      : GigabitEthernet0/0/2
  Pool-No        : 1
  Lease          : 1 Days 12 Hours 0 Minutes
  Domain-name    : -
  DNS-server0    : 10.0.23.254
  NBNS-server0   : -
```

```
Netbios-type : -
Position       : Interface       Status        : Unlocked
Gateway-0      : 10.0.23.1
Network        : 10.0.23.0
Mask           : 255.255.255.0
VPN instance   : --
--------------------------------------------------------------------
Start          End         Total  Used  Idle(Expired)  Conflict  Disable
--------------------------------------------------------------------
10.0.23.1   10.0.23.254    253    1     251(0)          0         1
--------------------------------------------------------------------

<S2>display ip interface brief
···output omit···
Interface              IP Address/Mask       Physical      Protocol
MEth0/0/1              unassigned            down          down
NULL0                  unassigned            up            up(s)
Vlanif1                10.0.23.253/24        up            up
```

在上述回显信息中，深色部分表明 R1 从接口地址池中为客户端的 VLANIF 1 接口分配了 IP 地址。

关闭 S1 上 VLANIF 1 接口以清除接口现有的 IP 地址，然后重新开启此接口以便重新从 R3 的接口地址池中获取新的 IP 地址。

```
[S1]interface Vlanif 1
[S1-Vlanif1]shutdown
[S1-Vlanif1]undo shutdown
```

开启 R3 的 G0/0/1 接口，使 R3 可以通过此接口从接口地址池中分配 IP 地址。

```
[R3]interface GigabitEthernet 0/0/1
[R3-GigabitEthernet0/0/1]undo shutdown
```

验证 R3 从接口地址池中为 S1 的 VLANIF 1 接口分配了新的 IP 地址。

```
<R3>display ip pool interface GigabitEthernet 0/0/1
  Pool-name       : GigabitEthernet0/0/1
  Pool-No         : 1
  Lease           : 1 Days 12 Hours 0 Minutes
  Domain-name     : -
  DNS-server0     : 10.0.12.254
  NBNS-server0    : -
```

```
  Netbios-type    : -
  Position        : Interface           Status             : Unlocked
  Gateway-0       : 10.0.12.3
  Network         : 10.0.12.0
  Mask            : 255.255.255.0
  VPN instance    : --
 ----------------------------------------------------------------------
  Start         End          Total    Used    Idle(Expired)   Conflict   Disable
 ----------------------------------------------------------------------
  10.0.12.1     10.0.12.254   253      1       251(0)          0          1
 ----------------------------------------------------------------------

<S1>display ip interface brief
…output omit…
Interface              IP Address/Mask      Physical    Protocol
MEth0/0/1              unassigned           down        down
NULL0                  unassigned           up          up(s)
Vlanif1                10.0.12.253/24       up          up
```

注意：交换机获取地址后会自动生成一条指向DHCP服务器的缺省静态路由，详见如下配置文件。

任务验证

查看配置文件：

```
[R1]display current-configuration
[V200R007C00SPC600]
#
 sysname R1
#
dhcp enable
#
ip pool pool1
 gateway-list 10.0.12.1
 network 10.0.12.0 mask 255.255.255.0
 lease day 1 hour 12 minute 0
#
interface GigabitEthernet0/0/1
 shutdown
 ip address 10.0.12.1 255.255.255.0
 dhcp select global
```

```
#
interface GigabitEthernet0/0/2
 ip address 10.0.23.1 255.255.255.0
 dhcp select interface
 dhcp server excluded-ip-address 10.0.23.254
 dhcp server lease day 1 hour 12 minute 0
 dhcp server dns-list 10.0.23.254
#
user-interface con 0
 authentication-mode password
 set authentication password cipher %$%$+L'YR&IZt'4,)>-*#1H",}%K-oJ_M9+'1OU~bD(\WTqB}%N,%$%$user-interface vty 0 4
#
return

[R3]display current-configuration
[V200R007C00SPC600]
#
 sysname R3
#
dhcp enable
#
ip pool pool2
 gateway-list 10.0.23.3
 network 10.0.23.0 mask 255.255.255.0
 lease day 1 hour 12 minute 0
#
interface GigabitEthernet0/0/1
 ip address 10.0.12.3 255.255.255.0
 dhcp select interface
 dhcp server excluded-ip-address 10.0.12.254
 dhcp server lease day 1 hour 12 minute 0
 dhcp server dns-list 10.0.12.254
#
interface GigabitEthernet0/0/2
 shutdown
 ip address 10.0.23.3 255.255.255.0
 dhcp select global
#
user-interface con 0
```

```
 authentication-mode password
 set authentication password cipher %$%$ksXDMg7Ry6yUU:63:DQ),#/sQg"@*S\U#.s.bHW
xQ,y%#/v,%$%$
user-interface vty 0 4
#
return
```

```
<S1>display current-configuration
#
!Software Version V200R008C00SPC500
 sysname S1
#
 dhcp enable
#
interface Vlanif1
 ip address dhcp-alloc
#
 ip route-static 0.0.0.0 0.0.0.0 10.0.12.3
#
user-interface con 0
user-interface vty 0 4
#
return
```

```
<S2>display current-configuration
#
!Software Version V200R008C00SPC500
 sysname S2
#
 dhcp enable
#
interface Vlanif1
 ip address dhcp-alloc
#
 ip route-static 0.0.0.0 0.0.0.0 10.0.23.1
#
user-interface con 0
user-interface vty 0 4
#
return
```

项目小结

通过本项目的学习,同学们了解了路由器上 FTP 服务的配置、建立 FTP 连接的过程、FTP 服务器参数的配置以及与 FTP 服务器传输文件的方法。FTP 服务提供了在服务器和客户机之间文件的上传和下载功能。网络设备可以从 FTP 服务器获取 vrp 系统文件,也可以将日志文件和配置文件保存到 FTP 服务器进行备份,以方便设备的远程管理。通过本项目的学习,同学们还了解了 DHCP 全局地址池和接口地址池的配置方法以及在交换机端口启用 DHCP 发现功能和 IP 地址分配功能的方法。路由器的 DHCP 功能可以给联网设备自动分配 IP 地址,减少了管理员的工作量,避免了设备的 IP 地址冲突,也方便了客户端的配置。

项目 4 交换技术

项目目标

1. 掌握静态路由的配置和测试连通性的方法
2. 掌握通过配置缺省路由实现本地网络与外部网络间的访问
3. 掌握基于端口安全的网络配置
4. 掌握 VLAN 相关配置
5. 掌握配置和管理 STP 的方法

任务 1　交换网络基础

任务 1.1　基于静态的公司与分部互联

任务背景

基于静态的公司与分部互联

某公司有北京总部、上海分部和广州分部 3 个办公地点,各分部与总部之间使用路由器互联。北京、上海、广州的路由器分别为 R1、R2、R3,路由器需要配置静态路由,使所有计算机能够互相访问。网络拓扑如图 4-1 所示,具体要求如下:

(1) 路由器之间通过 VPN 互联;
(2) 某公司总部与分部之间通过静态路由互联;
(3) 计算机和路由器的 IP 和接口信息如图 4-1 所示。

任务规划

北京总部使用 192.168.1.0 网段,上海分部使用 172.16.1.0 网段,广州分部使用 10.10.10.0 网段,R1 与 R2 之间为 20.20.20.0 网段,R1 与 R3 之间为 30.30.30.0 网段,R2 与 R3 之间为 40.40.40.0 网段,所有网段均使用 24 位子网掩码。路由器配置相应的静态路由,使所有计算机均能互访。

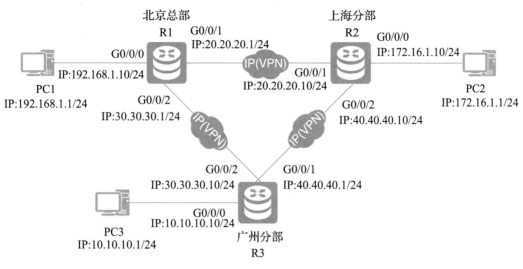

图 4-1 网络拓扑图

配置步骤如下：

(1) 配置路由器接口；

(2) 配置静态路由；

(3) 配置各计算机的 IP 地址。

具体规划见表 4-1、表 4-2、表 4-3。

表 4-1 IP 地址规划表

设备	接口	IP 地址
R1	G0/0/0	192.168.1.10/24
R1	G0/0/1	20.20.20.1/24
R1	G0/0/2	30.30.30.1/24
R2	G0/0/0	172.16.1.10/24
R2	G0/0/1	20.20.20.10/24
R2	G0/0/2	40.40.40.10/24
R3	G0/0/0	10.10.10.10/24
R3	G0/0/1	40.40.40.1/24
R3	G0/0/2	30.30.30.10/24
PC1	E0/0/1	192.168.1.1/24
PC2	E0/0/1	172.16.1.1/24
PC3	E0/0/1	10.10.10.1/24

表 4-2 端口规划表

本端设备	接口	端口 IP 地址	对端设备
R1	G0/0/0	192.168.1.10/24	PC1
R1	G0/0/1	20.20.20.1/24	R2
R1	G0/0/2	30.30.30.1/24	R3
R2	G0/0/0	172.16.1.10/24	PC2
R2	G0/0/1	20.20.20.10/24	R1
R2	G0/0/2	40.40.40.10/24	R3
R3	G0/0/0	10.10.10.10/24	PC3
R3	G0/0/1	40.40.40.1/24	R2
R3	G0/0/2	30.30.30.10/24	R1
PC1	E0/0/1	192.168.1.1/24	R1
PC2	E0/0/1	172.16.1.1/24	R2
PC3	E0/0/1	10.10.10.1/24	R3

表 4-3 路由规划表

路由器	目的网段	下一跳地址
R1	172.16.1.0/24	20.20.20.10
R1	10.10.10.0/24	30.30.30.10
R2	192.168.1.0/24	30.30.30.1
R2	10.10.10.0/24	40.40.40.1
R3	192.168.1.0/24	30.30.30.1
R3	172.16.1.0/24	40.40.40.10

任务实施

步骤一 配置路由器接口

(1) R1 的配置。

```
[Huawei]system-view
[Huawei]sysname R1
[R1]interface GigabitEthernet 0/0/0
[RGigabitEthernet0/0/0]ip address 192.168.1.10 255.255.255.0
[R1]interface GigabitEthernet 0/0/1
[RGigabitEthernet0/0/1]ip address 20.20.20.1 255.255.255.0
```

[R1]interface GigabitEthernet 0/0/2

[RGigabitEthernet0/0/2]ip address 30.30.30.1 255.255.255.0

（2）R2 的配置。

[Huawei]system-view

[Huawei]sysname R2

[R2]interface GigabitEthernet 0/0/0

[R2-GigabitEthernet0/0/0]ip address 172.16.1.10 255.255.255.0

[R2]interface GigabitEthernet 0/0/1

[R2-GigabitEthernet0/0/1]ip address 20.20.20.10 255.255.255.0

[R2]interface GigabitEthernet 0/0/2

[R2-GigabitEthernet0/0/2]ip address 40.40.40.10 255.255.255.0

（3）R3 的配置。

[Huawei]system-view

[Huawei]sysname R3

[R3]interface GigabitEthernet 0/0/0

[R3-GigabitEthernet0/0/0]ip address 10.10.10.10 255.255.255.0

[R3]interface GigabitEthernet 0/0/1

[R3-GigabitEthernet0/0/1]ip address 40.40.40.1 255.255.255.0

[R3]interface GigabitEthernet 0/0/2

[R3-GigabitEthernet0/0/2]ip address 30.30.30.10 255.255.255.0

步骤二 配置静态路由

在 R1 上配置目的网段为主机 PC2 所在网段的静态路由，即目的 IP 地址为 172.16.1.0，掩码 255.255.255.0 即 24 位。对于 R1 而言，要发送数据到主机 PC2，则必须先发送给 R2，所以 R2 即为 R1 的下一跳路由，R2 与 R1 所在的直连链路上的物理接口的 IP 地址即为下一跳 IP 地址，即 20.20.20.10。

[R1]ip route-static 172.16.1.0 24 20.20.20.10

配置目的网段为 PC3 所在网段的静态路由。

[R1]ip route-static 10.10.10.0 24 30.30.30.10

采取同样方式在 R2 上配置目的网段为 PC1、PC3 所在网段的静态路由。

[R2]ip route-static 192.168.1.0 24 20.20.20.1

[R2]ip route-static 10.10.10.0 24 40.40.40.1

采取同样方式在 R3 上配置目的网段为 PC1、PC2 所在网段的静态路由。

[R3]ip route-static 192.168.1.0 24 30.30.30.1

[R3]ip route-static 172.16.1.0 24 40.40.40.10

路由交换技术

步骤三 配置各计算机的 IP 地址

PC1、PC2、PC3 的 IP 地址配置分别如图 4-2、图 4-3、图 4-4 所示。

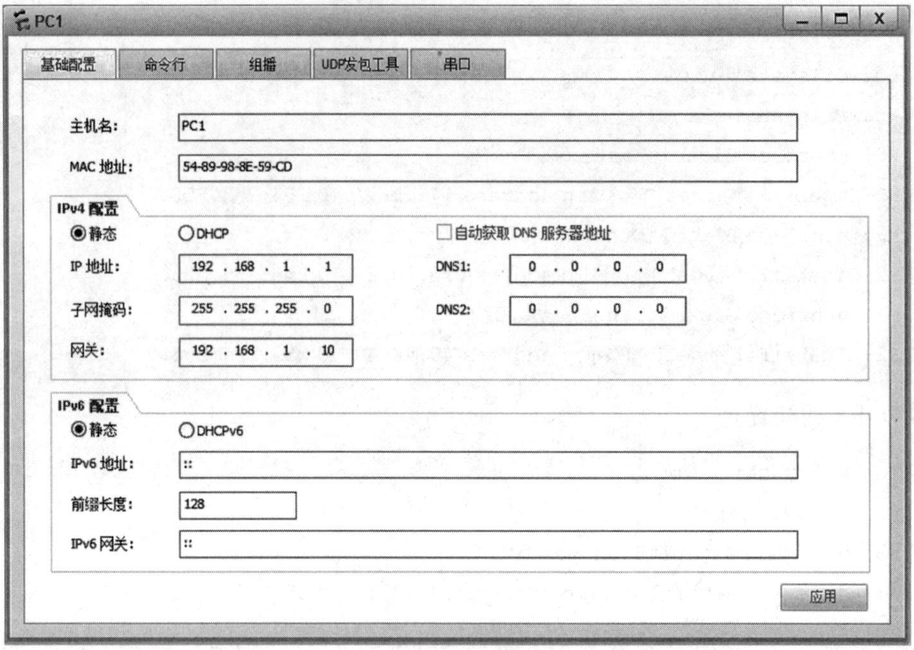

图 4-2 PC1 IP 地址配置图

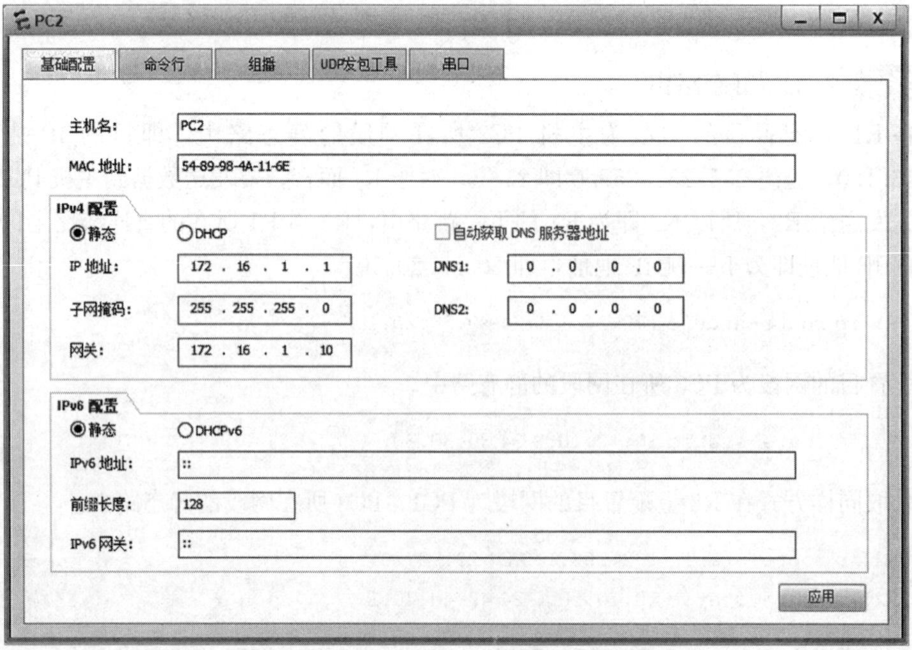

图 4-3 PC2 IP 地址配置图

图 4-4 PC3 IP 地址配置图

任务验证

步骤一 验证路由器上路由表的配置信息

（1）R1 的配置。

```
[R1]display ip routing-table
Route Flags: R-relay, D-download to fib
------------------------------------------------------------
Routing Tables: Public
        Destinations : 15        Routes : 15
Destination/Mask    Proto   Pre  Cost  Flags   NextHop         Interface
10.10.10.0/24       Static  60   0     RD      30.30.30.10     GigabitEthernet0/0/2
……
172.16.1.0/24       Static  60   0     RD      20.20.20.10     GigabitEthernet0/0/1
```

（2）R2 的配置。

```
[R2]display ip routing-table
Route Flags: R-relay, D-download to fib
------------------------------------------------------------
```

```
Routing Tables: Public
         Destinations : 15        Routes : 15
Destination/Mask  Proto  Pre  Cost  Flags  NextHop       Interface
10.10.10.0/24     Static  60   0    RD     40.40.40.1   GigabitEthernet0/0/2
……
192.168.1.0/24    Static  60   0    RD     20.20.20.1   GigabitEthernet0/0/1
```

（3）R3 的配置。

```
[R3]display ip routing-table
Route Flags: R - relay, D - download to fib
------------------------------------------------------------------------
Routing Tables: Public
         Destinations : 15        Routes : 15
Destination/Mask  Proto  Pre  Cost  Flags  NextHop       Interface
……
172.16.1.0/24     Static  60   0    RD     40.40.40.10  GigabitEthernet0/0/1
192.168.1.0/24    Static  60   0    RD     30.30.30.1   GigabitEthernet0/0/2
```

步骤二 测试各计算机的互通性

通过 ping 命令，测试各计算机内部通信情况。

使用 PC1 ping PC2 的计算机：

```
PC>ping 172.16.1.1
Ping 172.16.1.1: 32 data bytes, Press Ctrl_C to break
From 172.16.1.1: bytes=32 seq=1 ttl=126 time=15 ms
From 172.16.1.1: bytes=32 seq=2 ttl=126 time=16 ms
From 172.16.1.1: bytes=32 seq=3 ttl=126 time=16 ms
From 172.16.1.1: bytes=32 seq=4 ttl=126 time=15 ms
From 172.16.1.1: bytes=32 seq=5 ttl=126 time=16 ms

---172.16.1.1 ping statistics ---
  5 packet(s) transmitted
  5 packet(s) received
  0.00%  packet loss
  round-trip min/avg/max =15/15/16 ms
```

使用 PC1 ping PC3 的计算机：

```
PC>ping 10.10.10.1
Ping 10.10.10.1: 32 data bytes, Press Ctrl_C to break
From 10.10.10.1: bytes=32 seq=1 ttl=126 time=15 ms
From 10.10.10.1: bytes=32 seq=2 ttl=126 time=32 ms
```

```
From 10.10.10.1: bytes=32 seq=3 ttl=126 time=15 ms
From 10.10.10.1: bytes=32 seq=4 ttl=126 time=16 ms
From 10.10.10.1: bytes=32 seq=5 ttl=126 time=15 ms

---10.10.10.1 ping statistics ---
  5 packet(s) transmitted
  5 packet(s) received
  0.00%  packet loss
  round-trip min/avg/max =15/18/32 ms
```

可以看出，各计算机之间可以互相通信。

任务 1.2　基于端口安全的某公司网络组建

任务背景

某公司开发部为重要部门，所有员工使用指定的计算机工作，为防止员工或访客使用个人电脑接入网络，将使用基于端口安全策略组建开发部网络。网络拓扑图如图 4-5 所示，具体要求如下：

（1）开发部采用华为可网管交换机作为接入设备；

（2）出于安全的考虑，需要在交换机的端口上绑定指定计算机的 MAC 地址，以防止非法计算机的接入；

（3）计算机的 IP、MAC 和接入交换机的端口信息如图 4-5 所示。

基于端口安全的某公司网络组建

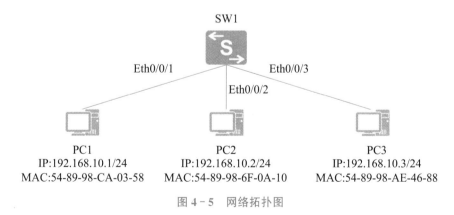

图 4-5　网络拓扑图

任务规划

MAC 地址是计算机的唯一物理标识，可以通过在交换机对应的端口上进行绑定，非绑定的 MAC 将无法接入网络中。查看 MAC 地址的方法有以下几种：

（1）在计算机中执行 ipconfig 命令即可查看本机的 MAC 地址；

（2）在计算机中执行 ARP-a 可以查看邻近计算机的 MAC 地址和 IP 地址；

(3) 在交换机上执行 display mac-address 命令可以查看对应端口上的 MAC 地址。

在进行端口绑定时，需要查看两个信息：一是计算机的 MAC 地址，二是计算机接入的端口。因此，我们可以先从计算机上查看本机的 MAC 地址，然后从交换机上查看 MAC 地址对应的端口，最后进行 MAC 地址和端口的绑定。

配置步骤如下：
(1) 查看计算机本地 MAC 地址；
(2) 查看 MAC 所在的交换机端口；
(3) 开启该交换机端口的端口安全，并绑定对应的 MAC 地址。

项目规划见表 4-4、表 4-5。

表 4-4 端口规划表

本端设备	端口号	对端设备
SW1	Eth0/0/1	PC1
SW1	Eth0/0/2	PC2
SW1	Eth0/0/3	PC3

表 4-5 IP 地址规划表

计算机	IP 地址	MAC 地址
PC1	192.168.10.1/24	54-89-98-CA-03-58
PC2	192.168.10.2/24	54-89-98-6F-0A-10
PC3	192.168.10.3/24	54-89-98-AE-46-88

任务实施

步骤一 查看计算机本地 MAC 地址

配置好计算机 IP 地址，在计算机命令行下输入 ipconfig，查看 MAC 地址。
(1) PC1 的配置。

```
PC>ipconfig

Link local IPv6 address.........: fe80::5689:98ff:feca:358
IPv6 address....................: :: / 128
IPv6 gateway....................: ::
IPv4 address....................: 192.168.10.1
Subnet mask.....................: 255.255.255.0
Gateway.........................: 0.0.0.0
Physical address................: 54-89-98-CA-03-58
DNS server......................:
```

（2）PC2 的配置。

```
PC>ipconfig

Link local IPv6 address..........: fe80::5689:98ff:fe6f:a10
IPv6 address.....................: :: / 128
IPv6 gateway.....................: ::
IPv4 address.....................: 192.168.10.2
Subnet mask.....................: 255.255.255.0
Gateway.........................: 0.0.0.0
Physical address................: 54-89-98-6F-0A-10
DNS server......................:
```

（3）PC3 的配置。

```
PC>ipconfig

Link local IPv6 address..........: fe80::5689:98ff:feae:4688
IPv6 address.....................: :: / 128
IPv6 gateway.....................: ::
IPv4 address.....................: 192.168.10.3
Subnet mask.....................: 255.255.255.0
Gateway.........................: 0.0.0.0
Physical address................: 54-89-98-AE-46-88
DNS server......................:
```

步骤二 查看 MAC 所在的交换机端口

在交换机上使用命令 display mac-address，查看交换机与计算机之间连接的端口对应的 MAC 地址。

```
<Huawei>system-view
[Huawei]sysname SW1
[SW1]
[SW1]display mac-address
MAC address table of slot 0:
------------------------------------------------------------------------
MAC Address  VLAN/ PEVLAN CEVLAN Port    Type     LSP/LSR-ID   VSI/SI  MAC-Tunnel
------------------------------------------------------------------------
5489-98ca-0358 1    -      -     Eth0/0/1  dynamic   0/-
5489-986f-0a10 1    -      -     Eth0/0/2  dynamic   0/-
5489-98ae-4688 1    -      -     Eth0/0/3  dynamic   0/-
------------------------------------------------------------------------
Total matching items on slot 0 displayed = 3
```

步骤三 开启该交换机端口的端口安全功能，并绑定对应的 MAC 地址

在交换机端口上打开端口安全功能，将 MAC 地址绑定到相对应接口上，并在 VLAN 1 上有效。

```
[SW1]interface Eth0/0/1
[SW1-Ethernet0/0/1]port-security enable
[SW1-Ethernet0/0/1]port-security mac-address sticky
[SW1-Ethernet0/0/1]port-security mac-address sticky 5489-98ca-0358 vlan 1

[SW1]interface Eth0/0/2
[SW1-Ethernet0/0/2]port-security enable
[SW1-Ethernet0/0/2]port-security mac-address sticky
[SW1-Ethernet0/0/2]port-security mac-address sticky 5489-986f-0a10 vlan 1

[SW1]interface Eth0/0/3
[SW1-Ethernet0/0/3]port-security enable
[SW1-Ethernet0/0/3]port-security mac-address sticky
[SW1-Ethernet0/0/3]port-security mac-address sticky 5489-98ae-4688 vlan 1
```

任务验证

步骤一 在交换机上查看配置是否生效

在交换机上使用 display mac-address 命令，查看交换机与计算机之间连接的端口的类型是否变为 sticky。

```
[SW1]display mac-address
MAC address table of slot 0:
------------------------------------------------------------------------
MAC Address  VLAN/ PEVLAN CEVLAN Port  Type   LSP/LSR-ID   VSI/SI  MAC-Tunnel
------------------------------------------------------------------------
5489-98ae-4688 1     -      -    Eth0/0/3    sticky        -
5489-98ca-0358 1     -      -    Eth0/0/1    sticky        -
5489-986f-0a10 1     -      -    Eth0/0/2    sticky        -
------------------------------------------------------------------------
Total matching items on slot 0 displayed = 3
```

步骤二 测试计算机的互通性

通过 ping 命令，测试内部通信情况。

使用 PC1 计算机 ping PC2 计算机：

```
PC>ping 192.168.10.2

Ping 192.168.10.2: 32 data bytes, Press Ctrl_C to break
From 192.168.10.2: bytes=32 seq=1 ttl=128 time=32 ms
From 192.168.10.2: bytes=32 seq=2 ttl=128 time=46 ms
From 192.168.10.2: bytes=32 seq=3 ttl=128 time=47 ms
From 192.168.10.2: bytes=32 seq=4 ttl=128 time=31 ms
From 192.168.10.2: bytes=32 seq=5 ttl=128 time=31 ms

---192.168.10.2 ping statistics ---
  5 packet(s) transmitted
  5 packet(s) received
  0.00% packet loss
  round-trip min/avg/max = 31/37/47 ms
```

使用 PC1 计算机 ping PC3 计算机：

```
PC>ping 192.168.10.3

Ping 192.168.10.3: 32 data bytes, Press Ctrl_C to break
From 192.168.10.3: bytes=32 seq=1 ttl=128 time=47 ms
From 192.168.10.3: bytes=32 seq=2 ttl=128 time=31 ms
From 192.168.10.3: bytes=32 seq=3 ttl=128 time=47 ms
From 192.168.10.3: bytes=32 seq=4 ttl=128 time=31 ms
From 192.168.10.3: bytes=32 seq=5 ttl=128 time=47 ms

---192.168.10.3 ping statistics ---
  5 packet(s) transmitted
  5 packet(s) received
  0.00% packet loss
  round-trip min/avg/max = 31/40/47 ms
```

可以看出，计算机之间可以互相通信。

步骤三　更换计算机，测试互通性

把计算机 PC3 更换为计算机 PC4，IP 地址相同，MAC 地址不同，连接到交换机 Eth0/0/3 接口上。

查看 PC4 的 MAC 地址：

```
PC>ipconfig

Link local IPv6 address..........: fe80::5689:98ff:fe87:617a
IPv6 address....................: :: / 128
IPv6 gateway....................: ::
IPv4 address....................: 192.168.10.3
Subnet mask.....................: 255.255.255.0
Gateway.........................: 0.0.0.0
Physical address................: 54-89-98-87-61-7A
DNS server......................:
```

使用 PC1 计算机 ping PC4 计算机：

```
PC>ping 192.168.10.3

Ping 192.168.10.3: 32 data bytes, Press Ctrl_C to break
From 192.168.10.1: Destination host unreachable
From 192.168.10.1: Destination host unreachable
From 192.168.10.1: Destination host unreachable
From 192.168.10.1: Destination host unreachable
From 192.168.10.1: Destination host unreachable
```

可以看出，更换计算机后，因为 MAC 地址不同，所以计算机之间不能通信。

任务 2　虚拟局域网技术

任务 2.1　为某公司创建部门 VLAN

任务背景

微课视频
为某公司创建部门 VLAN

某公司现有财务部、技术部和业务部，出于数据安全的考虑，各部门的计算机需要进行隔离。公司局域网使用一台 24 口二层交换机进行互联，其中：财务部的计算机 4 台，连接在 G0/0/1-4 端口；技术部的计算机 9 台，连接在 G0/0/5-12 及 G0/0/20 端口；业务部的计算机 8 台，连接在 G0/0/15-19 及 G0/0/21-23 端口。所有计算机采用 10.0.1.0/24 网段。现在在交换机中创建相应的 VLAN 以实现部门计算机的隔离。网络拓扑图如图 4-6 所示。

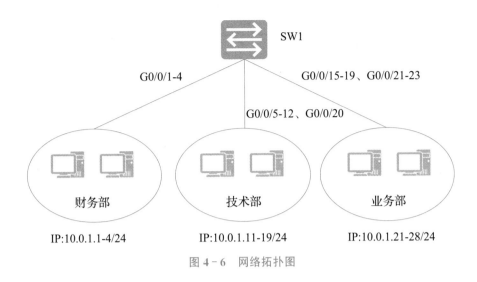

图 4-6 网络拓扑图

任务规划

默认情况下,二层交换机的所有端口都处于 VLAN 1 中。本项目中所有计算机均采用 10.0.1.0/24 网段,各计算机之间均可直接通信。为实现各部门之间的隔离,需要在交换机上创建 VLAN,并将各部门计算机的端口划分到相应的 VLAN 中。本项目将通过创建 VLAN 10、VLAN 20、VLAN 30 分别用于财务部、技术部、业务部内的计算机互联。

配置步骤如下:

(1) 创建 VLAN;

(2) 将端口划分至相应的 VLAN;

(3) 配置各部门计算机的 IP 地址。

项目规划见表 4-6、表 4-7、表 4-8。

表 4-6 VLAN 规划表

VLAN ID	IP 地址段	用途
VLAN 10	10.0.1.1-4/24	财务部
VLAN 20	10.0.1.11-19/24	技术部
VLAN 30	10.0.1.21-28/24	业务部

表 4-7 端口规划表

本端设备	端口号	端口类型	所属 VLAN	对端设备
SW1	G0/0/1-4	Access	VLAN 10	财务部 PC
SW1	G0/0/5-12、G0/0/20	Access	VLAN 20	技术部 PC
SW1	G0/0/15-19、G0/0/21-23	Access	VLAN 30	业务部 PC

表 4-8 IP 地址规划表

计算机	IP 地址
财务部-PC1	10.0.1.1/24
财务部-PC2	10.0.1.2/24
技术部-PC1	10.0.1.11/24
技术部-PC2	10.0.1.12/24
业务部-PC1	10.0.1.21/24
业务部-PC2	10.0.1.22/24

任务实施

步骤一 创建 VLAN

为各部门创建相应的 VLAN。

```
[Huawei]system-view
[Huawei]sysname SW1
[SW1]vlan 10
[SW1]vlan 20
[SW1]vlan 30
```

步骤二 将端口划分至相应的 VLAN

将各部门计算机所使用的端口按部门分别组成端口组，统一将端口类型转换为 Access 模式，并设置端口 PVID，将端口划分到相应的 VLAN。

```
[SW1]port-group group-member GigabitEthernet 0/0/1 to GigabitEthernet 0/0/4
[SW1-port-group]port link-type access
[SW1-GigabitEthernet0/0/1]port link-type access
[SW1-GigabitEthernet0/0/2]port link-type access
[SW1-GigabitEthernet0/0/3]port link-type access
[SW1-GigabitEthernet0/0/4]port link-type access

[SW1-port-group]port default vlan 10
[SW1-GigabitEthernet0/0/1]port default vlan 10
[SW1-GigabitEthernet0/0/2]port default vlan 10
[SW1-GigabitEthernet0/0/3]port default vlan 10
[SW1-GigabitEthernet0/0/4]port default vlan 10

[SW1]port-group group-member GigabitEthernet 0/0/5 to GigabitEthernet 0/0/12 GigabitEthernet 0/0/20
[SW1-port-group]port link-type access
[SW1-GigabitEthernet0/0/5]port link-type access
```

```
[SW1-GigabitEthernet0/0/6]port link-type access
[SW1-GigabitEthernet0/0/7]port link-type access
[SW1-GigabitEthernet0/0/8]port link-type access
[SW1-GigabitEthernet0/0/9]port link-type access
[SW1-GigabitEthernet0/0/10]port link-type access
[SW1-GigabitEthernet0/0/11]port link-type access
[SW1-GigabitEthernet0/0/12]port link-type access
[SW1-GigabitEthernet0/0/20]port link-type access

[SW1-port-group]port default vlan 20
[SW1-GigabitEthernet0/0/5]port default vlan 20
[SW1-GigabitEthernet0/0/6]port default vlan 20
[SW1-GigabitEthernet0/0/7]port default vlan 20
[SW1-GigabitEthernet0/0/8]port default vlan 20
[SW1-GigabitEthernet0/0/9]port default vlan 20
[SW1-GigabitEthernet0/0/10]port default vlan 20
[SW1-GigabitEthernet0/0/11]port default vlan 20
[SW1-GigabitEthernet0/0/12]port default vlan 20
[SW1-GigabitEthernet0/0/20]port default vlan 20

[SW1]port-group  group-member GigabitEthernet 0/0/15 to GigabitEthernet 0/0/19
GigabitEthernet 0/0/21 to GigabitEthernet 0/0/23
[SW1-port-group]port link-type access
[SW1-GigabitEthernet0/0/15]port link-type access
[SW1-GigabitEthernet0/0/16]port link-type access
[SW1-GigabitEthernet0/0/17]port link-type access
[SW1-GigabitEthernet0/0/18]port link-type access
[SW1-GigabitEthernet0/0/19]port link-type access
[SW1-GigabitEthernet0/0/21]port link-type access
[SW1-GigabitEthernet0/0/22]port link-type access
[SW1-GigabitEthernet0/0/23]port link-type access

[SW1-port-group]port default vlan 30
[SW1-GigabitEthernet0/0/5]port default vlan 30
[SW1-GigabitEthernet0/0/6]port default vlan 30
[SW1-GigabitEthernet0/0/7]port default vlan 30
[SW1-GigabitEthernet0/0/8]port default vlan 30
[SW1-GigabitEthernet0/0/9]port default vlan 30
[SW1-GigabitEthernet0/0/10]port default vlan 30
[SW1-GigabitEthernet0/0/11]port default vlan 30
[SW1-GigabitEthernet0/0/12]port default vlan 30
[SW1-GigabitEthernet0/0/20]port default vlan 30
```

步骤三 配置各部门计算机的 IP 地址

各部门计算机的 IP 地址配置分别如图 4-7、图 4-8、图 4-9 所示。

图 4-7 财务部-PC1 IP 地址配置图

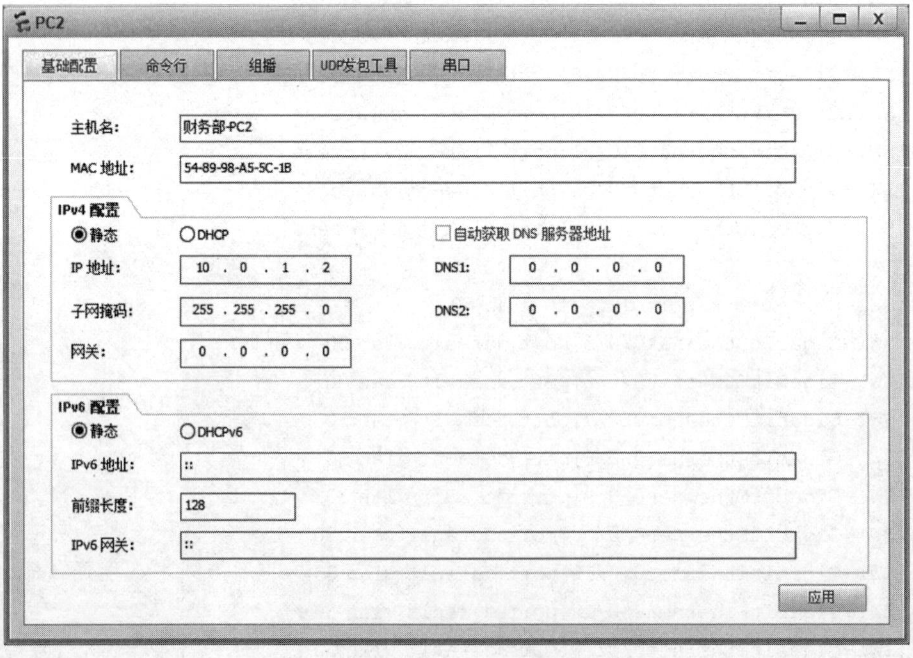

图 4-8 财务部-PC2 IP 地址配置图

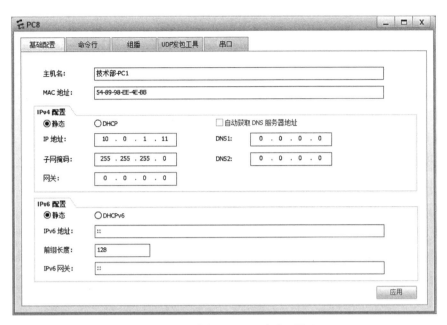

图 4-9 技术部-PC1 IP 地址配置图

任务验证

步骤一 验证交换机的 VLAN 配置信息

```
[SW1]display vlan
The total number of vlans is : 4
--------------------------------------------------------------------------------
U: Up;         D: Down;         TG: Tagged;         UT: Untagged;
MP: Vlan-mapping;                ST: Vlan-stacking;
#: ProtocolTransparent-vlan;     *: Management-vlan;
--------------------------------------------------------------------------------
VID  Type    Ports
--------------------------------------------------------------------------------
1    common  UT:GE0/0/13(D)   GE0/0/14(D)   GE0/0/15(U)   GE0/0/16(U)
                GE0/0/17(D)   GE0/0/18(D)   GE0/0/19(D)   GE0/0/21(D)
                GE0/0/22(D)   GE0/0/23(D)   GE0/0/24(D)

10   common  UT:GE0/0/1(U)    GE0/0/2(U)    GE0/0/3(U)    GE0/0/4(U)

20   common
30   common  UT:GE0/0/5(U)    GE0/0/6(U)    GE0/0/7(U)    GE0/0/8(D)

                GE0/0/9(D)    GE0/0/10(D)   GE0/0/11(D)   GE0/0/12(D)
                GE0/0/20(D)
```

```
VID   Status   Property       MAC-LRN  Statistics  Description
--------------------------------------------------------------------------------
1     enable   default        enable   disable     VLAN 0001
10    enable   default        enable   disable     VLAN 0010
20    enable   default        enable   disable     VLAN 0020
30    enable   default        enable   disable     VLAN 0030
```

步骤二 测试各部门计算机的互通性

通过 ping 命令，测试各部门内部通信情况。

使用财务部的计算机 ping 本部门的计算机：

```
PC>ping 10.0.1.2
Ping 10.0.1.2: 32 data bytes, Press Ctrl_C to break
From 10.0.1.2: bytes=32 seq=1 ttl=128 time=47 ms
From 10.0.1.2: bytes=32 seq=2 ttl=128 time=31 ms
From 10.0.1.2: bytes=32 seq=3 ttl=128 time=31 ms
From 10.0.1.2: bytes=32 seq=4 ttl=128 time=16 ms
From 10.0.1.2: bytes=32 seq=5 ttl=128 time=31 ms

---10.0.1.2 ping statistics ---
  5 packet(s) transmitted
  5 packet(s) received
  0.00%packet loss
  round-trip min/avg/max =16/31/47 ms
```

使用财务部的计算机 ping 技术部的计算机：

```
PC>ping 10.0.1.11
Ping 10.0.1.11: 32 data bytes, Press Ctrl_C to break
From 10.0.1.1: Destination host unreachable
From 10.0.1.1: Destination host unreachable
From 10.0.1.1: Destination host unreachable
From 10.0.1.1: Destination host unreachable
From 10.0.1.1: Destination host unreachable

--- 10.0.1.11 ping statistics ---
  5 packet(s) transmitted
  0 packet(s) received
  100.00%packet loss
```

可以看出，将端口加入不同的 VLAN 后，相同 VLAN 中的计算机可以互相通信，不同 VLAN 中的计算机则不可以互相通信。

任务 2.2 跨交换机的 VLAN

任务背景

某公司现有财务部和技术部，出于数据安全的考虑，各部门的计算机需要进行隔离。公司办公地点有两层楼，通过两台 24 口二层交换机进行互联，两台交换机均通过 G0/0/1 互联。财务部和技术部在这两层楼均有员工办公，其中财务部计算机使用 SW1 的 Eth0/0/1-5 端口及 SW2 的 Eth0/0/1-5 端口，技术部计算机使用 SW1 的 Eth0/0/6-10 及 SW2 的 Eth0/0/6-10 端口，所有计算机采用 10.0.1.0/24 网段。现在在各交换机中创建相应的 VLAN 以实现跨交换机的 VLAN 内通信，以及部门间计算机的相互隔离。网络拓扑图如图 4-10 所示。

微课视频

跨交换机的 VLAN

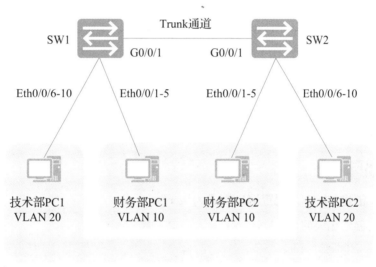

图 4-10 网络拓扑图

任务规划

为实现各部门之间的隔离，需要在交换机中创建 VLAN，并将各部门计算机的相应端口划分到相关的 VLAN 中，其中 VLAN 10、VLAN 20 分别用于财务部、技术部。同时，因为同一个 VLAN 中的计算机分属在不同交换机上，故级联的通道应配置为 Trunk 类型，使其能传输不同 VLAN 的数据帧。

配置步骤如下：

（1）创建 VLAN，配置端口模式为 Access；

（2）配置端口模式为 Trunk；

（3）配置各部门计算机的 IP 地址。

规划见表 4-9、表 4-10、表 4-11。

表 4-9　VLAN 规划表

VLAN ID	IP 地址段	用途
VLAN 10	10.0.1.1-10/24	财务部
VLAN 20	10.0.1.11-20/24	技术部

表 4-10　端口规划表

本端设备	端口号	端口类型	所属 VLAN	对端设备
SW1	Eth0/0/1-5	Access	VLAN 10	财务部 PC1
SW1	Eth0/0/6-10	Access	VLAN 20	技术部 PC1
SW1	G0/0/1	Trunk		SW2
SW2	Eth0/0/1-5	Access	VLAN 10	财务部 PC2
SW2	Eth0/0/6-10	Access	VLAN 20	技术部 PC2
SW2	G0/0/1	Trunk		SW1

表 4-11　IP 地址规划表

计算机	IP 地址
财务部 PC1	10.0.1.1/24
财务部 PC2	10.0.1.5/24
技术部 PC1	10.0.1.11/24
技术部 PC2	10.0.1.20/24

任务实施

步骤一　创建 VLAN，配置端口模式为 Access

为各部门创建相应的 VLAN。

（1）SW1 的配置。

```
[Huawei]system-view
[Huawei]sysname SW1
[SW1]vlan 10
[SW1-vlan10]description Fiance
[SW1]vlan 20
[SW1-vlan20]description Technical
```

（2）SW2 的配置。

```
[Huawei]system-view
[Huawei]sysname SW2
[SW2]vlan 10
[SW2-vlan10]description Fiance
[SW2]vlan 20
[SW2-vlan20]description Technical
```

将各部门计算机所使用的端口按部门分别组成端口组，统一将端口类型转换为 Access 模式，并设置端口 PVID，将端口划分到相应的 VLAN。

(1) SW1 的配置。

```
[SW1]port-group group-member Ethernet 0/0/1 to Ethernet 0/0/5
[SW1-port-group]port link-type access
[SW1-Ethernet0/0/1]port link-type access
[SW1-Ethernet0/0/2]port link-type access
[SW1-Ethernet0/0/3]port link-type access
[SW1-Ethernet0/0/4]port link-type access
[SW1-Ethernet0/0/5]port link-type access
[SW1-port-group]port default vlan 10
[SW1-Ethernet0/0/1]port default vlan 10
[SW1-Ethernet0/0/2]port default vlan 10
[SW1-Ethernet0/0/3]port default vlan 10
[SW1-Ethernet0/0/4]port default vlan 10
[SW1-Ethernet0/0/5]port default vlan 10
[SW1-port-group]quit
[SW1]port-group group-member Ethernet 0/0/6 to Ethernet 0/0/10
[SW1-port-group]port link-type access
[SW1-Ethernet0/0/6]port link-type access
[SW1-Ethernet0/0/7]port link-type access
[SW1-Ethernet0/0/8]port link-type access
[SW1-Ethernet0/0/9]port link-type access
[SW1-Ethernet0/0/10]port link-type access
[SW1-port-group]port default vlan 20
[SW1-Ethernet0/0/6]port default vlan 20
[SW1-Ethernet0/0/7]port default vlan 20
[SW1-Ethernet0/0/8]port default vlan 20
[SW1-Ethernet0/0/9]port default vlan 20
[SW1-Ethernet0/0/10]port default vlan 20
[SW1-port-group]quit
```

(2) SW2 的配置。

```
[SW2]port-group group-member Ethernet 0/0/1 to Ethernet 0/0/5
[SW2-port-group]port link-type access
[SW2-Ethernet0/0/1]port link-type access
[SW2-Ethernet0/0/2]port link-type access
[SW2-Ethernet0/0/3]port link-type access
[SW2-Ethernet0/0/4]port link-type access
[SW2-Ethernet0/0/5]port link-type access
```

```
[SW2-port-group]port default vlan 10
[SW2-Ethernet0/0/1]port default vlan 10
[SW2-Ethernet0/0/2]port default vlan 10
[SW2-Ethernet0/0/3]port default vlan 10
[SW2-Ethernet0/0/4]port default vlan 10
[SW2-Ethernet0/0/5]port default vlan 10
[SW2-port-group]quit
[SW2]port-group group-member Ethernet 0/0/6 to Ethernet 0/0/10
[SW2-port-group]port link-type access
[SW2-Ethernet0/0/6]port link-type access
[SW2-Ethernet0/0/7]port link-type access
[SW2-Ethernet0/0/8]port link-type access
[SW2-Ethernet0/0/9]port link-type access
[SW2-Ethernet0/0/10]port link-type access
[SW2-port-group]port default vlan 20
[SW2-Ethernet0/0/6]port default vlan 20
[SW2-Ethernet0/0/7]port default vlan 20
[SW2-Ethernet0/0/8]port default vlan 20
[SW2-Ethernet0/0/9]port default vlan 20
[SW2-Ethernet0/0/10]port default vlan 20
[SW2-port-group]quit
```

配置完成后，使用 display port vlan 命令检查 VLAN 和端口配置情况。

（1）SW1 的配置。

```
[SW1]display port vlan
Port                 Link Type    PVID    Trunk VLAN List
-------------------------------------------------------------
Ethernet0/0/1        access       10      -
Ethernet0/0/2        access       10      -
Ethernet0/0/3        access       10      -
Ethernet0/0/4        access       10      -
Ethernet0/0/5        access       10      -
Ethernet0/0/6        access       20      -
Ethernet0/0/7        access       20      -
Ethernet0/0/8        access       20      -
Ethernet0/0/9        access       20      -
Ethernet0/0/10       access       20      -
Ethernet0/0/11       hybrid       1       -
Ethernet0/0/12       hybrid       1       -
省略部分内容……
```

(2) SW2 的配置。

```
[SW2]display port vlan
Port                    Link Type    PVID    Trunk VLAN List
-------------------------------------------------------------------------
Ethernet0/0/1           access       10      -
Ethernet0/0/2           access       10      -
Ethernet0/0/3           access       10      -
Ethernet0/0/4           access       10      -
Ethernet0/0/5           access       10      -
Ethernet0/0/6           access       20      -
Ethernet0/0/7           access       20      -
Ethernet0/0/8           access       20      -
Ethernet0/0/9           access       20      -
Ethernet0/0/10          access       20      -
Ethernet0/0/11          hybrid       1       -
Ethernet0/0/12          hybrid       1       -
省略部分内容……
```

步骤二 配置端口模式为 Trunk

在 SW1 上配置 G0/0/1 为 Trunk 端口，允许 VLAN 10 和 VLAN 20 通过。

```
[SW1]interface GigabitEthernet 0/0/1
[SW1-GigabitEthernet0/0/1]port link-type trunk
[SW1-GigabitEthernet0/0/1]port trunk allow-pass vlan 10 20
```

在 SW2 上配置 G0/0/1 为 Trunk 端口，允许 VLAN 10 和 VLAN 20 通过。

```
[SW2]interface GigabitEthernet 0/0/1
[SW2-GigabitEthernet0/0/1]port link-type trunk
[SW2-GigabitEthernet0/0/1]port trunk allow-pass vlan 10 20
```

配置完成后，使用 display port vlan GigabitEthernet 0/0/1 命令检查 GE0/0/1 端口配置情况。

```
[SW1]display port vlan GigabitEthernet 0/0/1
Port                    Link Type    PVID    Trunk VLAN List
-------------------------------------------------------------------------
GigabitEthernet0/0/1    trunk        1       1  10  20
```

步骤三 配置各部门计算机的 IP 地址

各部门计算机的 IP 地址配置分别如图 4-11～图 4-14 所示。

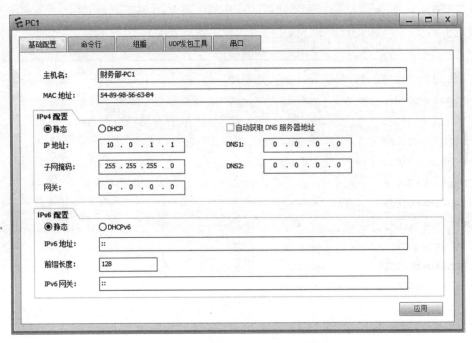

图 4-11 财务部-PC1 IP 地址配置图

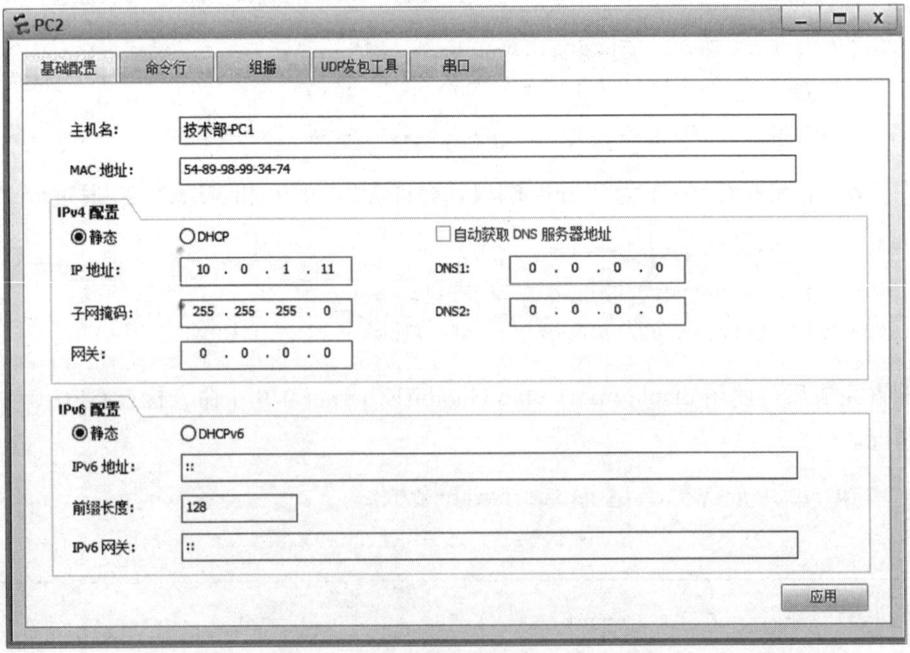

图 4-12 技术部-PC1 IP 地址配置图

图 4‑13　财务部-PC2 IP 地址配置图

图 4‑14　技术部-PC2 IP 地址配置图

任务验证

步骤一　**验证 SW1 交换机的 VLAN 配置信息**

在交换机上使用命令 display vlan，查看交换机详细的 VLAN 信息。

```
[SW1]display vlan
The total number of vlans is : 3
--------------------------------------------------------------------------------
U: Up;         D: Down;          TG: Tagged;          UT: Untagged;
MP: Vlan-mapping;               ST: Vlan-stacking;
#: ProtocolTransparent-vlan;    *: Management-vlan;
--------------------------------------------------------------------------------
VID  Type    Ports
--------------------------------------------------------------------------------
1    common  UT:Eth0/0/11(D)  Eth0/0/12(D)   Eth0/0/13(D)   Eth0/0/14(D)
                Eth0/0/15(D)  Eth0/0/16(D)   Eth0/0/17(D)   Eth0/0/18(D)
                Eth0/0/19(D)  Eth0/0/20(D)   Eth0/0/21(D)   Eth0/0/22(D)
                GE0/0/1(U)    GE0/0/2(D)

10   common  UT:Eth0/0/1(U)   Eth0/0/2(D)    Eth0/0/3(D)    Eth0/0/4(D)
                Eth0/0/5(D)
             TG:GE0/0/1(U)

20   common  UT:Eth0/0/6(D)   Eth0/0/7(D)    Eth0/0/8(D)    Eth0/0/9(D)
                Eth0/0/10(U)
             TG:GE0/0/1(U)


VID  Status  Property    MAC-LRN Statistics Description
--------------------------------------------------------------------------------
1    enable  default     enable  disable    VLAN 0001
10   enable  default     enable  disable    Fiance
20   enable  default     enable  disable    Technical
```

步骤二 验证 SW2 交换机的 VLAN 配置信息

```
[SW2]display vlan
The total number of vlans is : 3
--------------------------------------------------------------------------------
U: Up;         D: Down;          TG: Tagged;          UT: Untagged;
MP: Vlan-mapping;               ST: Vlan-stacking;
#: ProtocolTransparent-vlan;    *: Management-vlan;
--------------------------------------------------------------------------------
```

```
VID  Type    Ports
--------------------------------------------------------------------------
1    common  UT:Eth0/0/11(D)   Eth0/0/12(D)   Eth0/0/13(D)   Eth0/0/14(D)
             Eth0/0/15(D)      Eth0/0/16(D)   Eth0/0/17(D)   Eth0/0/18(D)
             Eth0/0/19(D)      Eth0/0/20(D)   Eth0/0/21(D)   Eth0/0/22(D)
             GE0/0/1(U)        GE0/0/2(D)

10   common  UT:Eth0/0/1(U)    Eth0/0/2(D)    Eth0/0/3(D)    Eth0/0/4(D)
             Eth0/0/5(D)
             TG:GE0/0/1(U)

20   common  UT:Eth0/0/6(D)    Eth0/0/7(D)    Eth0/0/8(D)    Eth0/0/9(D)
             Eth0/0/10(U)
             TG:GE0/0/1(U)


VID  Status  Property   MAC-LRN  Statistics  Description
--------------------------------------------------------------------------
1    enable  default    enable   disable     VLAN 0001
10   enable  default    enable   disable     Fiance
20   enable  default    enable   disable     Technical
```

步骤三 测试各部门计算机的互通性

通过 ping 命令，测试各部门内部通信情况。

使用财务部的计算机 ping 本部门的计算机：

```
PC>ping 10.0.1.5
Ping 10.0.1.5: 32 data bytes, Press Ctrl_C to break
From 10.0.1.5: bytes=32 seq=1 ttl=128 time=78 ms
From 10.0.1.5: bytes=32 seq=2 ttl=128 time=93 ms
From 10.0.1.5: bytes=32 seq=3 ttl=128 time=110 ms
From 10.0.1.5: bytes=32 seq=4 ttl=128 time=110 ms
From 10.0.1.5: bytes=32 seq=5 ttl=128 time=125 ms
---10.0.1.5 ping statistics ---
  5 packet(s) transmitted
  5 packet(s) received
  0.00%packet loss
  round-trip min/avg/max =78/103/125 ms
```

使用财务部的计算机 ping 技术部的计算机：

```
PC>ping 10.0.1.11
Ping 10.0.1.11: 32 data bytes, Press Ctrl_C to break
From 10.0.1.1: Destination host unreachable
From 10.0.1.1: Destination host unreachable
From 10.0.1.1: Destination host unreachable
From 10.0.1.1: Destination host unreachable
From 10.0.1.1: Destination host unreachable
--- 10.0.1.11 ping statistics ---
  5 packet(s) transmitted
  0 packet(s) received
  100.00%packet loss
```

可以看出，将端口加入不同的 VLAN 后，相同 VLAN 中的计算机可以互相通信，不同 VLAN 中的计算机则不可以互相通信。

任务3 基于 STP 的可靠网络配置

任务背景

某公司为提高网络的可靠性，使用了两台高性能交换机作为核心交换机，接入层交换机与核心层交换机互联，形成冗余结构，网络拓扑图如图 4-15 所示，具体要求如下：

（1）为避免交换环路问题，需要配置交换机的 STP 功能，要求核心交换机有较高优先级，SW1 为根交换机，SW2 为备用根交换机，SW1-SW3 和 SW1-SW4 为主链路；

（2）技术部使用 VLAN 10，网络地址为 10.0.1/24，PC1 和 PC2 分别接入 SW3 和 SW4。

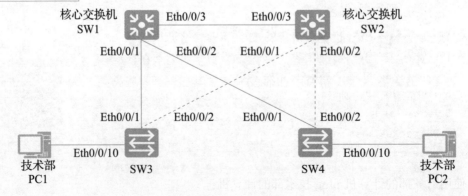

图 4-15 网络拓扑图

📋 任务规划

根据网络拓扑图 4-15 可知，SW1 和 SW2 为核心交换机，其中将 SW1 配置为根交换机，SW2 为备用根交换机；SW3 和 SW4 为接入交换机，其中 SW1-SW3 和 SW1-SW4 的链路为主链路，SW2-SW4 和 SW2-SW3 的链路为备用链路。

因此在 STP 配置中可将 SW1 的优先级设为最高，SW2 的优先级为次高。例如：SW1 的优先级为 0，SW2 的优先级为 4096。

同时，考虑到技术部的计算机划分在 VLAN 10 的网段内，且计算机连接在不同的交换机上，故交换机之间的链路需配置为 Trunk 模式。

具体配置步骤如下：

（1）创建 VLAN；
（2）将交换机端口划分至相应 VLAN；
（3）开启 STP；
（4）配置 STP 优先级；
（5）配置各部门计算机的 IP 地址。

具体规划见表 4-12、表 4-13、表 4-14。

表 4-12 VLAN 规划表

VLAN ID	VLAN 描述信息	IP 地址段	用途
VLAN 10	Technical	10.0.1.1-5/24	技术部

表 4-13 端口规划表

本端设备	端口号	端口类型	所属 VLAN	对端设备
SW1	Eth0/0/1	Trunk		SW3
SW1	Eth0/0/2	Trunk		SW4
SW1	Eth0/0/3	Trunk		SW2
SW2	Eth0/0/1	Trunk		SW3
SW2	Eth0/0/2	Trunk		SW4
SW2	Eth0/0/3	Trunk		SW1
SW3	Eth0/0/1	Trunk		SW1
SW3	Eth0/0/2	Trunk		SW2
SW3	Eth0/0/10	Access	VLAN 10	技术部 PC1
SW4	Eth0/0/1	Trunk		SW1
SW4	Eth0/0/2	Trunk		SW2
SW4	Eth0/0/10	Access	VLAN 10	技术部 PC2

表 4-14　IP 地址规划表

计算机	IP 地址
技术部 PC1	10.0.1.1/24
技术部 PC2	10.0.1.2/24

任务实施

步骤一　创建 VLAN

为各部门创建相应的 VLAN。

（1）SW1 的配置。

```
[Huawei]system-view
[Huawei]sysname SW1
[SW1]vlan 10
[SW1-vlan10]description Technical
```

（2）SW2 的配置。

```
[Huawei]system-view
[Huawei]sysname SW2
[SW2]vlan 10
[SW2-vlan10]description Technical
```

（3）SW3 的配置。

```
[Huawei]system-view
[Huawei]sysname SW3
[SW3]vlan 10
[SW3-vlan10]description Technical
```

（4）SW4 的配置。

```
[Huawei]system-view
[Huawei]sysname SW4
[SW4]vlan 10
[SW4-vlan10]description Technical
```

步骤二　将交换机端口划分至相应 VLAN

（1）SW1 的配置。

```
[SW1]port-group group-member Ethernet 0/0/1 to Ethernet 0/0/3
[SW1-port-group]port link-type trunk
[SW1-Ethernet0/0/1]port link-type trunk
[SW1-Ethernet0/0/2]port link-type trunk
```

[SW1-Ethernet0/0/3]port link-type trunk

[SW1-port-group]port trunk allow-pass vlan 10

[SW1-Ethernet0/0/1]port trunk allow-pass vlan 10

[SW1-Ethernet0/0/2]port trunk allow-pass vlan 10

[SW1-Ethernet0/0/3]port trunk allow-pass vlan 10

（2）SW2 的配置。

[SW2]port-group group-member Ethernet 0/0/1 to Ethernet 0/0/3

[SW2-port-group]port link-type trunk

[SW2-Ethernet0/0/1]port link-type trunk

[SW2-Ethernet0/0/2]port link-type trunk

[SW2-Ethernet0/0/3]port link-type trunk

[SW2-port-group]port trunk allow-pass vlan 10

[SW2-Ethernet0/0/1]port trunk allow-pass vlan 10

[SW2-Ethernet0/0/2]port trunk allow-pass vlan 10

[SW2-Ethernet0/0/3]port trunk allow-pass vlan 10

（3）SW3 的配置。

[SW3]interface Ethernet0/0/10

[SW3-Ethernet0/0/10]port link-type access

[SW3-Ethernet0/0/10]port default vlan 10

[SW3]port-group group-member Ethernet 0/0/1 to Ethernet 0/0/2

[SW3-port-group]port link-type trunk

[SW3-Ethernet0/0/1]port link-type trunk

[SW3-Ethernet0/0/2]port link-type trunk

[SW3-port-group]port trunk allow-pass vlan 10

[SW3-Ethernet0/0/1]port trunk allow-pass vlan 10

[SW3-Ethernet0/0/2]port trunk allow-pass vlan 10

（4）SW4 的配置。

[SW4]interface Ethernet0/0/10

[SW4-Ethernet0/0/10]port link-type access

[SW4-Ethernet0/0/10]port default vlan 10

[SW4]port-group group-member Ethernet 0/0/1 to Ethernet 0/0/2

[SW4-port-group]port link-type trunk

[SW4-Ethernet0/0/1]port link-type trunk

[SW4-Ethernet0/0/2]port link-type trunk

[SW4-port-group]port trunk allow-pass vlan 10

[SW4-Ethernet0/0/1]port trunk allow-pass vlan 10

[SW4-Ethernet0/0/2]port trunk allow-pass vlan 10

步骤三 开启 STP

(1) SW1 的配置。

```
[SW1]stp enable
[SW1]stp mode stp
```

(2) SW2 的配置。

```
[SW2]stp enable
[SW2]stp mode stp
```

(3) SW3 的配置。

```
[SW3]stp enable
[SW3]stp mode stp
```

(4) SW4 的配置。

```
[SW4]stp enable
[SW4]stp mode stp
```

步骤四 配置 STP 优先级

将 SW1 配置为根交换机，将 SW2 配置为备份根交换机。

方法 1：SW1 的优先级改为 0，SW2 的优先级改为 4096。

(1) SW1 的配置。

```
[SW1]stp priority 0
```

(2) SW2 的配置。

```
[SW2]stp priority 4096
```

方法 2：首先删除在 SW1 上所配置的优先级，使用 stp root primary 命令配置根交换机。

```
[SW1]undo stp priority
[SW1]stp root primary
```

删除在 SW2 上所配置的优先级，使用 stp root secondary 命令配置备份根交换机。

```
[SW2]undo stp priority
[SW2]stp root secondary
```

步骤五 配置技术部计算机的 IP 地址

技术部计算机的 IP 地址配置如图 4-16、图 4-17 所示。

图 4-16 技术部-PC1 IP 地址配置图

图 4-17 技术部-PC2 IP 地址配置图

任务验证

步骤一　验证各交换机的 VLAN 配置信息

（1）SW1 的配置。

```
[SW1]dis vlan
The total number of vlans is : 2
--------------------------------------------------------------------
U: Up;          D: Down;        TG: Tagged;         UT: Untagged;
MP: Vlan-mapping;               ST: Vlan-stacking;
#: ProtocolTransparent-vlan;    *: Management-vlan;
--------------------------------------------------------------------
VID  Type    Ports
--------------------------------------------------------------------
1    common  UT:Eth0/0/1(U)    Eth0/0/2(U)    Eth0/0/3(U)    Eth0/0/4(D)
                Eth0/0/5(D)    Eth0/0/6(D)    Eth0/0/7(D)    Eth0/0/8(D)
                Eth0/0/9(D)    Eth0/0/10(D)   Eth0/0/11(D)   Eth0/0/12(D)
                Eth0/0/13(D)   Eth0/0/14(D)   Eth0/0/15(D)   Eth0/0/16(D)
                Eth0/0/17(D)   Eth0/0/18(D)   Eth0/0/19(D)   Eth0/0/20(D)
                Eth0/0/21(D)   Eth0/0/22(D)   GE0/0/1(D)     GE0/0/2(D)
10   common  TG:Eth0/0/1(U)                   Eth0/0/2(U)    Eth0/0/3(U)
VID  Status  Property       MAC-LRN Statistics Description
--------------------------------------------------------------------
1    enable  default        enable  disable    VLAN 0001
10   enable  default        enable  disable    Technical
```

（2）SW2 的配置。

```
[SW2]dis vlan
The total number of vlans is : 2
--------------------------------------------------------------------
U: Up;          D: Down;        TG: Tagged;         UT: Untagged;
MP: Vlan-mapping;               ST: Vlan-stacking;
#: ProtocolTransparent-vlan;    *: Management-vlan;
--------------------------------------------------------------------
VID  Type    Ports
--------------------------------------------------------------------
1    common  UT:Eth0/0/1(U)    Eth0/0/2(U)    Eth0/0/3(U)    Eth0/0/4(D)
                Eth0/0/5(D)    Eth0/0/6(D)    Eth0/0/7(D)    Eth0/0/8(D)
                Eth0/0/9(D)    Eth0/0/10(D)   Eth0/0/11(D)   Eth0/0/12(D)
                Eth0/0/13(D)   Eth0/0/14(D)   Eth0/0/15(D)   Eth0/0/16(D)
                Eth0/0/17(D)   Eth0/0/18(D)   Eth0/0/19(D)   Eth0/0/20(D)
                Eth0/0/21(D)   Eth0/0/22(D)   GE0/0/1(D)     GE0/0/2(D)
10   common  TG:Eth0/0/1(U)    Eth0/0/2(U)    Eth0/0/3(U)
```

```
VID  Status  Property         MAC-LRN Statistics  Description
--------------------------------------------------------------------------------
1    enable  default          enable  disable     VLAN 0001
10   enable  default          enable  disable     Technical
```

(3) SW3 的配置。

```
[SW3]dis vlan
The total number of vlans is : 2
--------------------------------------------------------------------------------
U: Up;          D: Down;           TG: Tagged;         UT: Untagged;
MP: Vlan-mapping;                  ST: Vlan-stacking;
#: ProtocolTransparent-vlan;       *: Management-vlan;
--------------------------------------------------------------------------------
VID  Type    Ports
--------------------------------------------------------------------------------
1    common  UT:Eth0/0/1(U)    Eth0/0/2(U)    Eth0/0/3(D)    Eth0/0/4(D)
                Eth0/0/5(D)    Eth0/0/6(D)    Eth0/0/7(D)    Eth0/0/8(D)
                Eth0/0/9(D)    Eth0/0/11(D)   Eth0/0/12(D)   Eth0/0/13(D)
                Eth0/0/14(D)   Eth0/0/15(D)   Eth0/0/16(D)   Eth0/0/17(D)
                Eth0/0/18(D)   Eth0/0/19(D)   Eth0/0/20(D)   Eth0/0/21(D)
                Eth0/0/22(D)   GE0/0/1(D)     GE0/0/2(D)
10   common  UT:Eth0/0/10(U)
             TG:Eth0/0/1(U)    Eth0/0/2(U)
VID  Status  Property         MAC-LRN Statistics  Description
--------------------------------------------------------------------------------
1    enable  default          enable  disable     VLAN 0001
10   enable  default          enable  disable     Technical
```

(4) SW4 的配置。

```
[SW4]dis vlan
The total number of vlans is : 2
--------------------------------------------------------------------------------
U: Up;          D: Down;           TG: Tagged;         UT: Untagged;
MP: Vlan-mapping;                  ST: Vlan-stacking;
#: ProtocolTransparent-vlan;       *: Management-vlan;
--------------------------------------------------------------------------------
VID  Type    Ports
--------------------------------------------------------------------------------
1    common  UT:Eth0/0/1(U)    Eth0/0/2(U)    Eth0/0/3(D)    Eth0/0/4(D)
                Eth0/0/5(D)    Eth0/0/6(D)    Eth0/0/7(D)    Eth0/0/8(D)
```

```
                    Eth0/0/9(D)     Eth0/0/11(D)    Eth0/0/12(D)    Eth0/0/13(D)
                    Eth0/0/14(D)    Eth0/0/15(D)    Eth0/0/16(D)    Eth0/0/17(D)
                    Eth0/0/18(D)    Eth0/0/19(D)    Eth0/0/20(D)    Eth0/0/21(D)
                    Eth0/0/22(D)    GE0/0/1(D)      GE0/0/2(D)
10   common  UT:Eth0/0/10(U)
              TG:Eth0/0/1(U)    Eth0/0/2(U)
VID  Status  Property     MAC-LRN Statistics Description
--------------------------------------------------------------------------
1    enable  default      enable  disable     VLAN 0001
10   enable  default      enable  disable     Technical
```

步骤二 查看各交换机的 STP 状态

查看各交换机的 STP 状态信息，SW1、SW2 使用 display stp 命令查看 stp 模式是否正确，SW3、SW4 使用 display stp brief 命令查看备用端口是否处于 Discarding 状态。

（1）SW1 的配置。

```
[SW1]dis stp
-------[CIST Global Info][Mode STP]-------
CIST Bridge         :0    .4c1f-cc13-37a8
Config Times        :Hello 2s MaxAge 20s FwDly 15s MaxHop 20
Active Times        :Hello 2s MaxAge 20s FwDly 15s MaxHop 20
CIST Root/ERPC      :0    .4c1f-cc13-37a8 / 0
CIST RegRoot/IRPC   :0    .4c1f-cc13-37a8 / 0
……
```

（2）SW2 的配置。

```
[SW2]dis stp
-------[CIST Global Info][Mode STP]-------
CIST Bridge         :4096 .4c1f-ccf2-272d
Config Times        :Hello 2s MaxAge 20s FwDly 15s MaxHop 20
Active Times        :Hello 2s MaxAge 20s FwDly 15s MaxHop 20
CIST Root/ERPC      :0    .4c1f-cc13-37a8 / 200000
CIST RegRoot/IRPC   :4096 .4c1f-ccf2-272d / 0
……
```

（3）SW3 的配置。

```
[SW3]display stp brief
MSTID  Port              Role   STP State    Protection
  0    Ethernet0/0/1     ROOT   FORWARDING   NONE
  0    Ethernet0/0/2     ALTE   DISCARDING   NONE
  0    Ethernet0/0/10    DESI   FORWARDING   NONE
```

(4) SW4 的配置。

```
[SW4]display stp brief
MSTID  Port              Role   STP State    Protection
  0    Ethernet0/0/1     ROOT   FORWARDING   NONE
  0    Ethernet0/0/2     ALTE   DISCARDING   NONE
  0    Ethernet0/0/10    DESI   FORWARDING   NONE
```

步骤三 测试各部门计算机的互通性

通过 ping 命令，测试各部门内部通信情况。

```
PC>ping 10.0.1.2

Ping 10.0.1.2: 32 data bytes, Press Ctrl_C to break
From 10.0.1.2: bytes=32 seq=1 ttl=128 time=156 ms
From 10.0.1.2: bytes=32 seq=2 ttl=128 time=125 ms
From 10.0.1.2: bytes=32 seq=3 ttl=128 time=140 ms
From 10.0.1.2: bytes=32 seq=4 ttl=128 time=157 ms
From 10.0.1.2: bytes=32 seq=5 ttl=128 time=156 ms

---10.0.1.2 ping statistics ---
  5 packet(s) transmitted
  5 packet(s) received
  0.00%packet loss
  round-trip min/avg/max =125/146/157 ms
```

使用财务部的计算机 ping 本部门的计算机。

项目小结

　　随着互联网技术的发展，交换技术已成为网络的核心技术之一。通过本项目的学习，同学们理解了交换网络的基本概念，包括以太网的工作方式的概念；掌握了交换机的工作原理，通过项目操作实现交换机端口 MAC 地址绑定；掌握了虚拟局域网技术的基本原理，通过项目操作完成基本配置；掌握了生成树协议的基本原理，通过项目操作实现其基本应用。

项目 5 路由技术

项目目标

1. 掌握静态路由和浮动静态路由的配置方法
2. 掌握单臂路由的配置方法
3. 掌握 VLAN 间路由的配置方法
4. 掌握 OSPF 的配置方法

任务 1 静态路由和浮动静态路由

任务背景

公司现有一个总部与两个分支机构。其中,R1 为总部路由器,R2、R3 为分支机构路由器,总部与分支机构间的路由器通过以太网实现互联,且当前公司网络中没有配置任何路由协议。要求实现全公司网络的互联互通。

任务规划

由于网络的规模比较小,因此拟配置通过静态路由和缺省路由来实现网络互通。IP 编址信息拓扑图如图 5-1 所示,IP 地址规划见表 5-1。

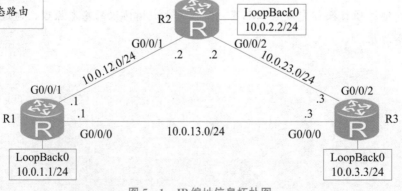

图 5-1 IP 编址信息拓扑图

表 5-1　IP 地址规划表

设备	接口	IP 地址
R1	LoopBack0	10.0.1.1/24
R1	G0/0/0	10.0.13.1/24
R1	G0/0/1	10.0.12.1/24
R2	LoopBack0	10.0.2.2/24
R2	G0/0/1	10.0.12.2/24
R2	G0/0/2	10.0.23.2/24
R3	LoopBack0	10.0.3.3/24
R3	G0/0/0	10.0.13.3/24
R3	G0/0/2	10.0.23.3/24

配置步骤如下：

（1）基础配置和 IP 编址；

（2）在 R2 上配置静态路由；

（3）查看 R2 的路由表；

（4）配置浮动静态路由；

（5）再次查看 R2 的路由表；

（6）查看数据的传输路径。

任务实施

步骤一　基础配置和 IP 编址

（1）在 R1 上配置设备名称和 IP 地址。

```
<Huawei>system-view
[Huawei]sysname R1
[R1]interface GigabitEthernet 0/0/0
[R1-GigabitEthernet0/0/0]ip address 10.0.13.1 24
[R1-GigabitEthernet0/0/0]quit
[R1]interface GigabitEthernet 0/0/1
[R1-GigabitEthernet0/0/1]ip address 10.0.12.1 24
[R1-GigabitEthernet0/0/1]quit
[R1]interface LoopBack 0
[R1-LoopBack0]ip address 10.0.1.1 24
```

(2) 在 R2 上配置设备名称和 IP 地址。

```
<Huawei>system-view
[Huawei]sysname R2
[R2]interface GigabitEthernet 0/0/1
[R2-GigabitEthernet0/0/1]ip address 10.0.12.2 24
[R2-GigabitEthernet0/0/1]quit
[R2]interface GigabitEthernet0/0/2
[R2-GigabitEthernet0/0/2]ip add 10.0.23.2 24
[R2-GigabitEthernet0/0/2]quit
[R2]interface LoopBack0
[R2-LoopBack0]ip address 10.0.2.2 24
```

(3) 在 R3 上配置设备名称和 IP 地址。

```
<Huawei>system-view
[Huawei]sysname R3
[R3]interface GigabitEthernet 0/0/0
[R3-GigabitEthernet0/0/0]ip address 10.0.13.3 24
[R3-GigabitEthernet0/0/0]quit
[R3]interface GigabitEthernet0/0/2
[R3-GigabitEthernet0/0/2]ip address 10.0.23.3 24
[R3-GigabitEthernet0/0/2]quit
[R3]interface LoopBack 0
[R3-LoopBack0]ip address 10.0.3.3 24
```

步骤二 在 R2 上配置静态路由

配置目的地址为 10.0.13.0/24 和 10.0.3.0/24 的静态路由，路由的下一跳配置为 R3 的 G0/0/0 接口 IP 地址 10.0.23.3。默认静态路由优先级为 60，不需要额外配置路由优先级信息。

```
[R2]ip route-static 10.0.13.0 24 10.0.23.3
[R2]ip route-static 10.0.3.0 24 10.0.23.3
```

步骤三 查看 R2 的路由表

```
<R2>display ip routing-table
Route Flags: R - relay, D - download to fib
------------------------------------------------------------------------------
Routing Tables: Public
         Destinations : 15        Routes : 15
```

```
Destination/Mask      Proto   Pre  Cost  Flags  NextHop      Interface
     10.0.2.0/24      Direct  0    0     D      10.0.2.2     LoopBack0
     10.0.2.2/32      Direct  0    0     D      127.0.0.1    LoopBack0
     10.0.2.255/32    Direct  0    0     D      127.0.0.1    LoopBack0
     10.0.3.0/24      Static  60   0     RD     10.0.23.3    GigabitEthernet0/0/2
     10.0.12.0/24     Direct  0    0     D      10.0.12.2    GigabitEthernet0/0/1
     10.0.12.2/32     Direct  0    0     D      127.0.0.1    GigabitEthernet0/0/1
     10.0.12.255/32   Direct  0    0     D      127.0.0.1    GigabitEthernet0/0/1
     10.0.13.0/24     Static  60   0     RD     10.0.23.3    GigabitEthernet0/0/2
     10.0.23.0/24     Direct  0    0     D      10.0.23.2    GigabitEthernet0/0/2
     10.0.23.2/32     Direct  0    0     D      127.0.0.1    GigabitEthernet0/0/2
     10.0.23.255/32   Direct  0    0     D      127.0.0.1    GigabitEthernet0/0/2
     127.0.0.0/8      Direct  0    0     D      127.0.0.1    InLoopBack0
     127.0.0.1/32     Direct  0    0     D      127.0.0.1    InLoopBack0
     127.255.255.255/32 Direct 0   0     D      127.0.0.1    InLoopBack0
     255.255.255.255/32 Direct 0   0     D      127.0.0.1    InLoopBack0
```

步骤四 配置浮动静态路由

将备份静态路由的优先级修改为 80。当 R2 和 R3 间的链路发生故障时，使用备份静态路由传输数据，R2 还可以通过 R1 与 R3 通信。

```
[R1]ip route-static 10.0.3.0 24 10.0.13.3

[R2]ip route-static 10.0.13.0 255.255.255.0 10.0.12.1 preference 80
[R2]ip route-static 10.0.3.0 24 10.0.12.1 preference 80

[R3]ip route-static 10.0.12.0 24 10.0.13.1
```

步骤五 再次查看 R2 的路由表

```
<R2>display ip routing-table
Route Flags: R - relay, D - download to fib
------------------------------------------------------------------
Routing Tables: Public
         Destinations : 15      Routes : 15
Destination/Mask     Proto   Pre  Cost  Flags  NextHop      Interface
10.0.2.0/24          Direct  0    0     D      10.0.2.2     LoopBack0
10.0.2.2/32          Direct  0    0     D      127.0.0.1    LoopBack0
10.0.2.255/32        Direct  0    0     D      127.0.0.1    LoopBack0
10.0.3.0/24          Static  60   0     RD     10.0.23.3    GigabitEthernet0/0/2
10.0.12.0/24         Direct  0    0     D      10.0.12.2    GigabitEthernet0/0/1
```

10.0.12.2/32	Direct	0	0	D	127.0.0.1	GigabitEthernet0/0/1
10.0.12.255/32	Direct	0	0	D	127.0.0.1	GigabitEthernet0/0/1
10.0.13.0/24	Static	60	0	RD	10.0.23.3	GigabitEthernet0/0/2
10.0.23.0/24	Direct	0	0	D	10.0.23.2	GigabitEthernet0/0/2
10.0.23.2/32	Direct	0	0	D	127.0.0.1	GigabitEthernet0/0/2
10.0.23.255/32	Direct	0	0	D	127.0.0.1	GigabitEthernet0/0/2
127.0.0.0/8	Direct	0	0	D	127.0.0.1	InLoopBack0
127.0.0.1/32	Direct	0	0	D	127.0.0.1	InLoopBack0
127.255.255.255/32	Direct	0	0	D	127.0.0.1	InLoopBack0
255.255.255.255/32	Direct	0	0	D	127.0.0.1	InLoopBack0

步骤六 查看数据的传输路径

```
<R2>tracert 10.0.13.3
 traceroute to  10.0.13.3(10.0.13.3), max hops: 30 ,packet length: 40,
press CTRL_C to break
 1 10.0.23.3 40 ms   31 ms   30 ms

<R2>tracert 10.0.3.3
 traceroute to  10.0.3.3(10.0.3.3), max hops: 30 ,packet length: 40,
press CTRL_C to break
 1 10.0.23.3 40 ms   30 ms   30 ms
```

任务验证

验证浮动静态路由：

（1）关闭 R2 上的 G0/0/2 接口，模拟 R2 与 R3 间的链路发生故障。

```
[R2]interface GigabitEthernet 0/0/2
[R2-GigabitEthernet0/0/2]shutdown
[R2-GigabitEthernet0/0/2]quit
```

（2）查看 R2 的路由表。

```
<R2>display ip routing-table
Route Flags: R -relay, D -download to fib
------------------------------------------------------------------------------
Routing Tables: Public
        Destinations : 12    Routes : 12

Destination/Mask     Proto  Pre Cost Flags NextHop        Interface

       10.0.2.0/24   Direct  0   0    D    10.0.2.2       LoopBack0
       10.0.2.2/32   Direct  0   0    D    127.0.0.1      LoopBack0
```

10.0.2.255/32	Direct	0	0	D	127.0.0.1	LoopBack0
10.0.3.0/24	Static	80	0	RD	10.0.12.1	GigabitEthernet0/0/1
10.0.12.0/24	Direct	0	0	D	10.0.12.2	GigabitEthernet0/0/1
10.0.12.2/32	Direct	0	0	D	127.0.0.1	GigabitEthernet0/0/1
10.0.12.255/32	Direct	0	0	D	127.0.0.1	GigabitEthernet0/0/1
10.0.13.0/24	Static	80	0	RD	10.0.12.1	GigabitEthernet0/0/1
127.0.0.0/8	Direct	0	0	D	127.0.0.1	InLoopBack0
127.0.0.1/32	Direct	0	0	D	127.0.0.1	InLoopBack0
127.255.255.255/32	Direct	0	0	D	127.0.0.1	InLoopBack0
255.255.255.255/32	Direct	0	0	D	127.0.0.1	InLoopBack0

（3）执行 tracert 命令，查看数据包的转发路径。

```
<R2>tracert 10.0.13.3
 traceroute to 10.0.13.3(10.0.13.3), max hops: 30 ,packet length: 40,press CTRL_C
to break
  1 10.0.12.1 40 ms   21 ms   21 ms
  2 10.0.13.3 30 ms   21 ms   21 ms

<R2>tracert 10.0.3.3
 traceroute to 10.0.3.3(10.0.3.3), max hops: 30 ,packet length: 40,press CTRL_C
to break
  1 10.0.12.1 40 ms   21 ms   21 ms
  2 10.0.13.3 30 ms   21 ms   21 ms
```

任务 2　单臂路由配置

任务背景

某公司内部网络通常会通过划分不同的 VLAN 来隔离不同部门之间的二层通信，并保证各部门间的信息安全。但是由于业务需要，部分部门之间需要实现跨 VLAN 通信。

任务规划

网络管理员决定借助路由器，通过配置单臂路由来实现两个部门间的相互通信。

IP 编址信息拓扑图如图 5-2 所示。IP 地址规划见表 5-2。

微课视频

单臂路由配置

路由交换技术

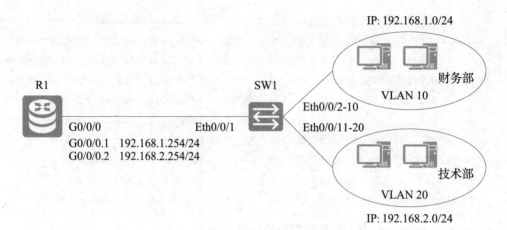

图 5-2 IP 编址信息拓扑图

表 5-2 IP 地址规划表

设备	接口	IP 地址
R1	G0/0/0.1	192.168.1.254/24
R1	G0/0/0.2	192.168.2.254/24
PC1	Eth0/0/1	192.168.1.1/24
PC2	Eth0/0/1	192.168.2.1/24

配置步骤如下：
(1) 配置交换机接口；
(2) 路由器配置；
(3) 配置各计算机的 IP 地址。

任务实施

步骤一 配置交换机接口

为各部门创建相应的 VLAN，将端口划分至相应 VLAN。

```
[Huawei]system-view
[Huawei]sysname SW1
[SW1]vlan batch 10 20
[SW1]int Ethernet0/0/2
[SW1-Ethernet0/0/2]port link-type access
[SW1-Ethernet0/0/2]port default vlan 10
[SW1-Ethernet0/0/2]quit
[SW1]interface Ethernet0/0/3
[SW1-Ethernet0/0/3]port link-type access
[SW1-Ethernet0/0/3]port default vlan 20
[SW1-Ethernet0/0/3]quit
```

```
[SW1]interface Ethernet0/0/1
[SW1-Ethernet0/0/1]port link-type trunk
[SW1]interface Ethernet0/0/1
[SW1-Ethernet0/0/1]port trunk allow-pass vlan 10 20
```

步骤二 路由器配置

在路由器以太网口上建立子接口，分别新建两个子接口，然后为两个子接口配置 IP 和掩码，作为 VLAN 的网关，同时启动 802.1Q 协议。

```
<Huawei>system-view
[Huawei]sysname R1
[R1]interface GigabitEthernet 0/0/0.1
[R1-GigabitEthernet0/0/0.1]dot1q termination vid 10
[R1-GigabitEthernet0/0/0.1]ip address 192.168.1.254 24
[R1-GigabitEthernet0/0/0.1]arp broadcast enable
[R1-GigabitEthernet0/0/0.1]quit
[R1]interface GigabitEthernet 0/0/0.2
[R1-GigabitEthernet0/0/0.2]dot1q termination vid 20
[R1-GigabitEthernet0/0/0.2]ip address 192.168.2.254 24
[R1-GigabitEthernet0/0/0.2]arp broadcast enable
```

步骤三 配置各计算机的 IP 地址

按照拓扑图规划配置两个部门主机的 IP 地址，分别如图 5-3、图 5-4 所示。

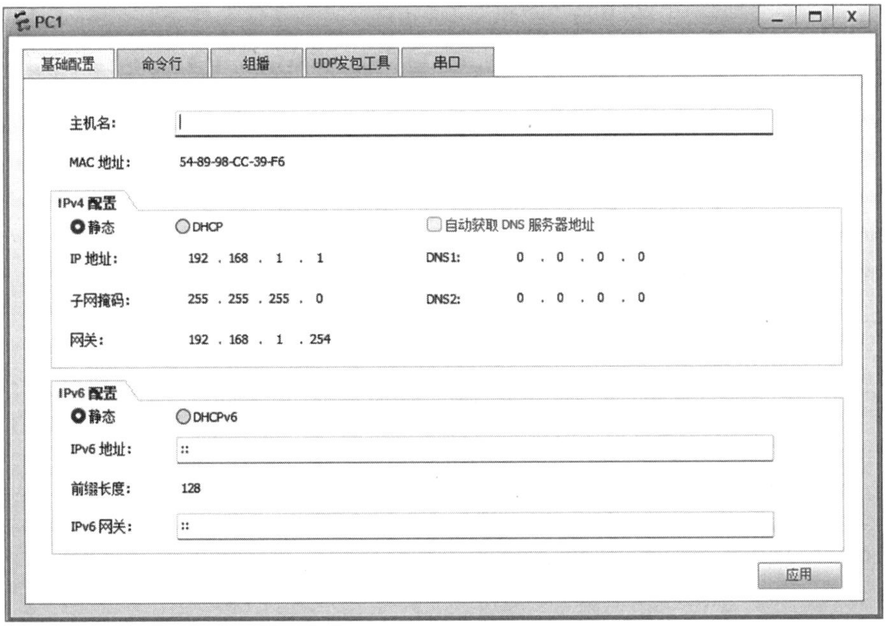

图 5-3 财务部 PC1 IP 地址配置图

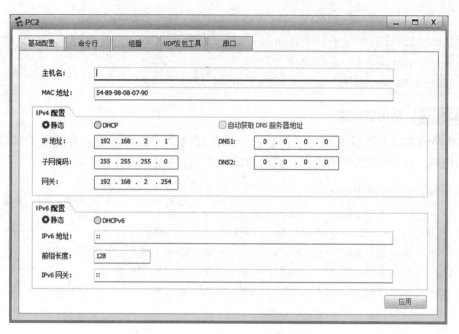

图 5-4 技术部 PC2 IP 地址配置图

任务验证

步骤一 查看路由表的配置信息

（1）使用 display ip interface brief 命令查看子接口的 IP 信息。

```
[R1]display ip interface brief
*down: administratively down
^down: standby
(l): loopback
(s): spoofing
The number of interface that is UP in Physical is 4
The number of interface that is DOWN in Physical is 2
The number of interface that is UP in Protocol is 3
The number of interface that is DOWN in Protocol is 3

Interface                  IP Address/Mask       Physical   Protocol
GigabitEthernet0/0/0       unassigned            up         down
GigabitEthernet0/0/0.1     192.168.1.254/24      up         up
GigabitEthernet0/0/0.2     192.168.2.254/24      up         up
GigabitEthernet0/0/1       unassigned            down       down
GigabitEthernet0/0/2       unassigned            down       down
NULL0
```

（2）使用 display ip routing-table 命令查看路由表的配置。

```
[R1]display ip routing-table
Route Flags: R-relay, D-download to fib
-------------------------------------------------------------------------------
Routing Tables: Public
         Destinations : 10        Routes : 10

Destination/Mask      Proto   Pre  Cost  Flags  NextHop         Interface

      127.0.0.0/8     Direct   0    0      D    127.0.0.1       InLoopBack0
      127.0.0.1/32    Direct   0    0      D    127.0.0.1       InLoopBack0
127.255.255.255/32    Direct   0    0      D    127.0.0.1       InLoopBack0
    192.168.1.0/24    Direct   0    0      D    192.168.1.254   GigabitEthernet0/0/0.1
  192.168.1.254/32    Direct   0    0      D    127.0.0.1       GigabitEthernet0/0/0.1
  192.168.1.255/32    Direct   0    0      D    127.0.0.1       GigabitEthernet0/0/0.1
    192.168.2.0/24    Direct   0    0      D    192.168.2.254   GigabitEthernet0/0/0.2
  192.168.2.254/32    Direct   0    0      D    127.0.0.1       GigabitEthernet0/0/0.2
  192.168.2.255/32    Direct   0    0      D    127.0.0.1       GigabitEthernet0/0/0.2
255.255.255.255/32    Direct   0    0      D    127.0.0.1       InLoopBack0
```

步骤二 测试各部门计算机的互通性

要求两个部门实现主机互通。通过 ping 命令，测试各部门内部通信情况，如图 5-5 所示。

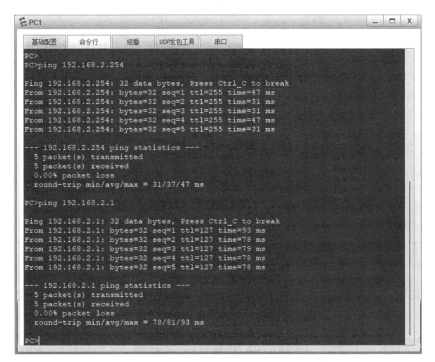

图 5-5 测试 PC 连通性 1

任务3 VLAN间路由配置

任务背景

VLAN间路由配置

某公司内部网络通常会通过划分不同的VLAN来隔离不同部门之间的二层通信,并保证各部门间的信息安全。但是由于业务需要,部分部门之间需要实现跨VLAN通信。

任务规划

三层交换机可以通过创建VLANIF的方式实现VLAN间的通信。在交换机中创建VLAN 10、VLAN 20分别用于财务部、技术部的计算机接入,VLAN 10使用192.168.1.0/24网段,VLAN 20使用192.168.2.0/24网段。在交换机中创建VLAN 10、VLAN 20的VLANIF接口,并配置对应的IP地址作为计算机的网关,即可实现VLAN间的通信。拓扑图如图5-6所示,IP地址规划见表5-3。

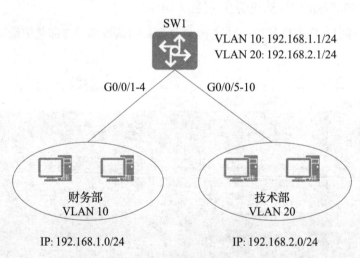

图5-6 IP编址信息拓扑图

表5-3 IP地址规划表

设备	接口	IP地址
SW1	VLANIF 10	192.168.1.1/24
SW1	VLANIF 10	192.168.2.1/24
PC1	E0/0/1	192.168.1.10/24
PC2	E0/0/1	192.168.2.20/24

配置步骤如下：
(1) 创建 VLAN；
(2) 将端口划分至相应的 VLAN；
(3) 配置 VLANIF 接口；
(4) 配置各计算机的 IP 地址。

任务实施

步骤一　创建 VLAN

为各部门创建相应的 VLAN。

```
[Huawei]system-view
[Huawei]sysname SW1
[SW1]vlan 10
[SW1]vlan 20
```

步骤二　将端口划分至相应的 VLAN

将各部门计算机所使用的端口类型转换为 Access 模式，并设置接口 PVID，将端口划分到相应的 VLAN。

```
[SW1]interface GigabitEthernet 0/0/1
[SW1-GigabitEthernet0/0/1]port link-type access
[SW1-GigabitEthernet0/0/1]port default vlan 10

[SW1]interface GigabitEthernet 0/0/5
[SW1-GigabitEthernet0/0/5]port link-type access
[SW1-GigabitEthernet0/0/5]port default vlan 20
```

步骤三　配置 VLANIF 接口

```
[SW1]interface Vlanif 10
[SW1-Vlanif10]ip add 192.168.1.1 24

[SW1]interface Vlanif 20
[SW1-Vlanif20]ip add 192.168.2.1 24
```

步骤四　配置各计算机的 IP 地址

按照拓扑图规划配置两个部门主机的 IP 地址，如图 5-7、图 5-8 所示。

路由交换技术

图 5-7　PC1 IP 地址配置图

图 5-8　PC2 IP 地址配置图

任务验证

步骤一　查看交换机的 VLAN 配置信息

（1）使用 display vlan 命令查看 VLAN 配置信息。

```
<SW1>display vlan
The total number of vlans is : 3
--------------------------------------------------------------------------------
U: Up;          D: Down;          TG: Tagged;          UT: Untagged;
MP: Vlan-mapping;               ST: Vlan-stacking;
#: ProtocolTransparent-vlan;    *: Management-vlan;
--------------------------------------------------------------------------------

VID  Type    Ports
--------------------------------------------------------------------------------
1    common  UT:GE0/0/2(D)    GE0/0/3(D)     GE0/0/4(D)     GE0/0/6(D)
                GE0/0/7(D)    GE0/0/8(D)     GE0/0/9(D)     GE0/0/10(D)
                GE0/0/11(D)   GE0/0/12(D)    GE0/0/13(D)    GE0/0/14(D)
                GE0/0/15(D)   GE0/0/16(D)    GE0/0/17(D)    GE0/0/18(D)
                GE0/0/19(D)   GE0/0/20(D)    GE0/0/21(D)    GE0/0/22(D)
                GE0/0/23(D)   GE0/0/24(D)

10   common  UT:GE0/0/1(U)

20   common  UT:GE0/0/5(U)


VID  Status  Property    MAC-LRN Statistics Description
--------------------------------------------------------------------------------

1    enable  default     enable  disable     VLAN 0001
10   enable  default     enable  disable     VLAN 0010
20   enable  default     enable  disable     VLAN 0020
```

(2) 使用 display ip interface brief 命令查看 IP 地址配置信息。

```
[SW1]display ip int brief
*down: administratively down
^down: standby
(l): loopback
(s): spoofing
The number of interface that is UP in Physical is 3
The number of interface that is DOWN in Physical is 2
The number of interface that is UP in Protocol is 3
The number of interface that is DOWN in Protocol is 2
```

```
Interface                    IP Address/Mask         Physical    Protocol
MEth0/0/1                    unassigned              down        down
NULL0                        unassigned              up          up(s)
Vlanif1                      unassigned              down        down
Vlanif10                     192.168.1.1/24          up          up
Vlanif20                     192.168.2.1/24          up          up
```

步骤二 测试各部门计算机的互通性

要求两个部门实现主机互通。通过 ping 命令，测试各部门内部通信情况，如图 5-9 所示。

图 5-9 测试 PC 连通性 2

任务 4　单区域 OSPF 配置

任务背景

单区域 OSPF 配置

公司有三台路由器实现网络连接，现需要配置路由协议，实现网络的互联互通。规划使用 OSPF 协议来进行路由信息的传递。采用单区域 OSPF 配置，网络中所有路由器均属于 OSPF 的区域 0。

任务规划

IP 编址信息拓扑图如图 5-10 所示，IP 地址规划见表 5-4。

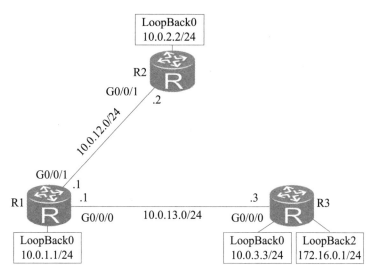

图 5-10 IP 编址信息拓扑图

表 5-4 IP 地址规划表

设备	接口	IP 地址
R1	LoopBack0	10.0.1.1/24
R1	G0/0/0	10.0.13.1/24
R1	G0/0/1	10.0.12.1/24
R2	LoopBack0	10.0.2.2/24
R2	G0/0/1	10.0.12.2/24
R3	LoopBack0	10.0.3.3/24
R3	LoopBack2	172.16.0.1/24
R3	G0/0/0	10.0.13.3/24

配置步骤如下：

（1）路由器基础配置；

（2）配置 OSPF。

任务实施

步骤一 路由器基础配置

完成路由器的基本配置以及 IP 编址。

（1）路由器 R1 的配置。

```
<Huawei>system-view
Enter system view, return user view with Ctrl+Z.
[Huawei]sysname R1
[R1]interface GigabitEthernet 0/0/1
```

```
[R1-GigabitEthernet0/0/1]ip address 10.0.12.1 24
[R1-GigabitEthernet0/0/1]quit
[R1]interface GigabitEthernet 0/0/0
[R1-GigabitEthernet0/0/0]ip address 10.0.13.1 24
[R1-GigabitEthernet0/0/0]quit
[R1]interface LoopBack 0
[R1-LoopBack0]ip address 10.0.1.1 24
```

（2）路由器 R2 的配置。

```
<Huawei>system-view
Enter system view, return user view with Ctrl+Z.
[Huawei]sysname R2
[R2]interface GigabitEthernet 0/0/1
[R2-GigabitEthernet0/0/1]ip address 10.0.12.2 24
[R2-GigabitEthernet0/0/1]quit
[R2]interface LoopBack 0
[R2-LoopBack0]ip address 10.0.2.2 24
```

（3）路由器 R3 的配置。

```
<Huawei>system-view
Enter system view, return user view with Ctrl+Z.
[Huawei]sysname R3
[R3]interface GigabitEthernet 0/0/0
[R3-GigabitEthernet0/0/0]ip address 10.0.13.3 24
[R3-GigabitEthernet0/0/0]quit
[R3]interface LoopBack 0
[R3-LoopBack0]ip address 10.0.3.3 24
[R3-LoopBack0]quit
[R3]interface LoopBack 2
[R3-LoopBack2]ip address 172.16.0.1 24
```

步骤二 配置 OSPF

（1）R1 的 Router ID 配置为 10.0.1.1（逻辑接口 LoopBack 0 的地址），开启 OSPF 进程 1（缺省进程），并将网段 10.0.1.0/24、10.0.12.0/24 和 10.0.13.0/24 发布到 OSPF 区域 0。

```
[R1]ospf 1 router-id 10.0.1.1
[R1-ospf-1]area 0
[R1-ospf-1-area-0.0.0.0]network 10.0.1.0 0.0.0.255
[R1-ospf-1-area-0.0.0.0]network 10.0.13.0 0.0.0.255
[R1-ospf-1-area-0.0.0.0]network 10.0.12.0 0.0.0.255
```

(2) 将 R2 的 Router ID 配置为 10.0.2.2，开启 OSPF 进程 1，并将网段 10.0.12.0/24 和 10.0.2.0/24 发布到 OSPF 区域 0。

```
[R2]ospf 1 router-id 10.0.2.2
[R2-ospf-1]area 0
[R2-ospf-1-area-0.0.0.0]network 10.0.2.0 0.0.0.255
[R2-ospf-1-area-0.0.0.0]network 10.0.12.0 0.0.0.255
```

(3) 将 R3 的 Router ID 配置为 10.0.3.3，开启 OSPF 进程 1，并将网段 10.0.3.0/24 和 10.0.13.0/24 发布到 OSPF 区域 0。

```
[R3]ospf 1 router-id 10.0.3.3
[R3-ospf-1]area 0
[R3-ospf-1-area-0.0.0.0]network 10.0.3.0 0.0.0.255
[R3-ospf-1-area-0.0.0.0]network 10.0.13.0 0.0.0.255
```

任务验证

查看路由配置信息：

(1) 使用 display ip routing-table 命令查看路由表。

路由器 R1 的路由表：

```
<R1>display ip routing-table
Route Flags: R-relay, D-download to fib
------------------------------------------------------------------------
Routing Tables: Public
         Destinations : 15     Routes : 15
Destination/Mask     Proto   Pre  Cost  Flags  NextHop        Interface
10.0.1.0/24          Direct  0    0     D      10.0.1.1       LoopBack0
10.0.1.1/32          Direct  0    0     D      127.0.0.1      LoopBack0
10.0.1.255/32        Direct  0    0     D      127.0.0.1      LoopBack0
10.0.2.2/32          OSPF    10   1     D      10.0.12.2      GigabitEthernet0/0/1
10.0.3.3/32          OSPF    10   1     D      10.0.13.3      GigabitEthernet0/0/0
10.0.12.0/24         Direct  0    0     D      10.0.12.1      GigabitEthernet0/0/1
10.0.12.1/32         Direct  0    0     D      127.0.0.1      GigabitEthernet0/0/1
10.0.12.255/32       Direct  0    0     D      127.0.0.1      GigabitEthernet0/0/1
10.0.13.0/24         Direct  0    0     D      10.0.13.1      GigabitEthernet0/0/0
10.0.13.1/32         Direct  0    0     D      127.0.0.1      GigabitEthernet0/0/0
10.0.13.255/32       Direct  0    0     D      127.0.0.1      GigabitEthernet0/0/0
127.0.0.0/8          Direct  0    0     D      127.0.0.1      InLoopBack0
127.0.0.1/32         Direct  0    0     D      127.0.0.1      InLoopBack0
127.255.255.255/32   Direct  0    0     D      127.0.0.1      InLoopBack0
255.255.255.255/32   Direct  0    0     D      127.0.0.1      InLoopBack0
```

路由器 R2 的路由表：

```
<R2>display ip routing-table
Route Flags: R-relay, D-download to fib
------------------------------------------------------------------------
Routing Tables: Public
        Destinations : 13      Routes : 13
Destination/Mask    Proto   Pre  Cost   Flags  NextHop      Interface
10.0.1.1/32         OSPF    10   1      D      10.0.12.1    GigabitEthernet0/0/1
10.0.2.0/24         Direct  0    0      D      10.0.2.2     LoopBack0
10.0.2.2/32         Direct  0    0      D      127.0.0.1    LoopBack0
10.0.2.255/32       Direct  0    0      D      127.0.0.1    LoopBack0
10.0.3.3/32         OSPF    10   2      D      10.0.12.1    GigabitEthernet0/0/1
10.0.12.0/24        Direct  0    0      D      10.0.12.2    GigabitEthernet0/0/1
10.0.12.2/32        Direct  0    0      D      127.0.0.1    GigabitEthernet0/0/1
10.0.12.255/32      Direct  0    0      D      127.0.0.1    GigabitEthernet0/0/1
10.0.13.0/24        OSPF    10   2      D      10.0.12.1    GigabitEthernet0/0/1
127.0.0.0/8         Direct  0    0      D      127.0.0.1    InLoopBack0
127.0.0.1/32        Direct  0    0      D      127.0.0.1    InLoopBack0
127.255.255.255/32  Direct  0    0      D      127.0.0.1    InLoopBack0
255.255.255.255/32  Direct  0    0      D      127.0.0.1    InLoopBack0
```

路由器 R3 的路由表：

```
<R3>display ip routing-table
Route Flags: R -relay, D -download to fib
------------------------------------------------------------------------
Routing Tables: Public
        Destinations : 16      Routes : 16
Destination/Mask    Proto   Pre  Cost   Flags  NextHop      Interface
10.0.1.1/32         OSPF    10   1      D      10.0.13.1    GigabitEthernet0/0/0
10.0.2.2/32         OSPF    10   2      D      10.0.13.1    GigabitEthernet0/0/0
10.0.3.0/24         Direct  0    0      D      10.0.3.3     LoopBack0
10.0.3.3/32         Direct  0    0      D      127.0.0.1    LoopBack0
10.0.3.255/32       Direct  0    0      D      127.0.0.1    LoopBack0
10.0.12.0/24        OSPF    10   2      D      10.0.13.1    GigabitEthernet0/0/0
10.0.13.0/24        Direct  0    0      D      10.0.13.3    GigabitEthernet0/0/0
10.0.13.3/32        Direct  0    0      D      127.0.0.1    GigabitEthernet0/0/0
10.0.13.255/32      Direct  0    0      D      127.0.0.1    GigabitEthernet0/0/0
127.0.0.0/8         Direct  0    0      D      127.0.0.1    InLoopBack0
127.0.0.1/32        Direct  0    0      D      127.0.0.1    InLoopBack0
127.255.255.255/32  Direct  0    0      D      127.0.0.1    InLoopBack0
```

172.16.0.0/24	Direct	0	0	D	172.16.0.1	LoopBack2
172.16.0.1/32	Direct	0	0	D	127.0.0.1	LoopBack2
172.16.0.255/32	Direct	0	0	D	127.0.0.1	LoopBack2
255.255.255.255/32	Direct	0	0	D	127.0.0.1	InLoopBack0

（2）检测 R2 和 R1 (10.0.1.1) 以及 R2 和 R3 (10.0.3.3) 间的连通性。

```
<R2>ping 10.0.1.1
  PING 10.0.1.1: 56  data bytes, press CTRL_C to break
    Reply from 10.0.1.1: bytes=56 Sequence=1 ttl=255 time=37 ms
    Reply from 10.0.1.1: bytes=56 Sequence=2 ttl=255 time=42 ms
    Reply from 10.0.1.1: bytes=56 Sequence=3 ttl=255 time=42 ms
    Reply from 10.0.1.1: bytes=56 Sequence=4 ttl=255 time=45 ms
    Reply from 10.0.1.1: bytes=56 Sequence=5 ttl=255 time=42 ms
  ---10.0.1.1 ping statistics ---
    5 packet(s) transmitted
    5 packet(s) received
    0.00%packet loss
    round-trip min/avg/max = 37/41/45 ms
<R2>ping 10.0.3.3
  PING 10.0.3.3: 56  data bytes, press CTRL_C to break
    Reply from 10.0.3.3: bytes=56 Sequence=1 ttl=254 time=37 ms
    Reply from 10.0.3.3: bytes=56 Sequence=2 ttl=254 time=42 ms
    Reply from 10.0.3.3: bytes=56 Sequence=3 ttl=254 time=42 ms
    Reply from 10.0.3.3: bytes=56 Sequence=4 ttl=254 time=42 ms
    Reply from 10.0.3.3: bytes=56 Sequence=5 ttl=254 time=42 ms

  ---10.0.3.3 ping statistics ---
    5 packet(s) transmitted
    5 packet(s) received
    0.00%packet loss
round-trip min/avg/max = 37/41/42 ms
```

（3）执行 display ospf peer 命令，查看 OSPF 邻居状态。

```
<R1>display ospf peer
         OSPF Process 1 with Router ID 10.0.1.1
                Neighbors
Area 0.0.0.0 interface 10.0.12.1(GigabitEthernet0/0/1)'s neighbors
Router ID: 10.0.2.2        Address: 10.0.12.2
  State: Full  Mode:Nbr is  Master  Priority: 1
  DR: 10.0.12.1  BDR: 10.0.12.2  MTU: 0
```

```
        Dead timer due in 32   sec
        Retrans timer interval: 5
        Neighbor is up for 00:47:59
        Authentication Sequence: [ 0 ]
                   Neighbors
 Area 0.0.0.0 interface 10.0.13.1(GigabitEthernet0/0/0)'s neighbors
 Router ID: 10.0.3.3         Address: 10.0.13.3
        State: Full   Mode:Nbr is  Master  Priority: 1
        DR: 10.0.13.1  BDR: 10.0.13.3  MTU: 0
        Dead timer due in 34   sec
        Retrans timer interval: 5
        Neighbor is up for 00:41:44
        Authentication Sequence: [ 0 ]
```

项目小结

本项目介绍了静态路由、浮动静态路由、VLAN 间路由和单区域 OSPF 路由的配置。通过本项目的学习，同学们掌握了静态路由的配置方式、浮动静态路由的原理和实现；掌握了使用单臂路由和三层交换机实现 VLAN 间通信的方法；掌握了单区域 OSPF 路由的配置方式。本项目还介绍了查看路由表的方法，以及执行 tracert 命令查看数据包的转发路径的方法。

项目 6 网络可靠性

项目目标

1. 掌握基于 VRRP 的 ISP 双出口备份链路配置方法
2. 掌握基于 VRRP 的负载均衡出口链路的配置方法
3. 掌握链路聚合的配置方法

任务 1　基于 VRRP 的 ISP 双出口备份链路配置

任务背景

某公司原采用 ISP-A 作为接入服务商,用于内部计算机访问互联网的出口。为提高接入互联网的可靠性,该公司现增加 ISP-B 作为备用接入服务商,当 ISP-A 的接入链路出现故障时,启用 ISP-B 的接入链路。公司网络拓扑图如图 6-1 所示,项目具体要求如下:

微课视频

基于 VRRP 的 ISP 双出口备份链路配置

(1) 路由器 AR1 和 AR2 通过拨号方式接入 ISP 网络 AR3 和 AR4,AR5 是互联网的一台路由器;

(2) 网络出口以 ISP-A 作为主链路,ISP-B 作为备份链路,当主链路失败时,自动启用备份链路,以保证内部网络与互联网服务器的连通。

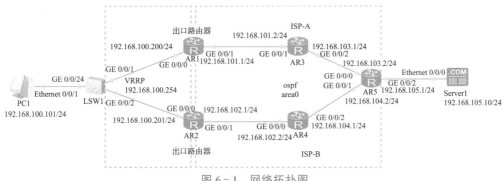

图 6-1　网络拓扑图

(3) 路由器间采用 OSPF 动态路由协议互联；

(4) 拓扑测试计算机和路由器的 IP 地址及接口信息如图 6-1 所示。

任务规划

AR1、AR2 为连接服务商 ISP-A、ISP-B 的出口路由器，其中 AR1 为主路由器，AR2 为备份路由器。为实现出口路由器的主备自动切换，首先需要在 AR1、AR2 上启用 VRRP 功能，设置虚拟网关 192.168.100.254，并将 AR1 的优先级设至为 110，即优先级最高，AR2 的优先级为默认的 100，此时 AR1 为主路由器；其次，配置对 AR1 路由器 GE 0/0/1 接口的链路状态跟踪，当链路状态为 DOWN 时，AR1 的 VRRP 优先级下降至 50，此时 AR2 切换为主路由器。内部计算机方面，在连接到网络后，将默认网关指向 VRRP 虚拟网关 192.168.100.254，此时计算机的出口链路会根据 VRRP 的状态选择主路由器作为出口。在互联网连接方面，由于 ISP-A、ISP-B 均采用 OSPF 协议，故所有路由器均配置 OSPF 协议，并设置为 Area0 区域。

配置步骤如下：

(1) 配置路由器接口；

(2) 部署 OSPF 协议；

(3) 配置 VRRP 协议；

(4) 配置上行接口监视；

(5) 配置各部门计算机的 IP 地址。

IP 地址规划见表 6-1。

表 6-1 IP 地址规划表

设备	接口	IP 地址
AR1	GE 0/0/0	192.168.100.200/24
AR1	GE 0/0/1	192.168.101.1/24
AR2	GE 0/0/0	192.168.100.201/24
AR2	GE 0/0/1	192.168.102.1/24
AR3	GE 0/0/1	192.168.101.2/24
AR3	GE 0/0/2	192.168.103.1/24
AR4	GE 0/0/0	192.168.102.2/24
AR4	GE 0/0/2	192.168.104.1/24
AR5	GE 0/0/0	192.168.103.2/24
AR5	GE 0/0/1	192.168.104.2/24
AR5	GE 0/0/2	192.168.105.1/24

任务实施

步骤一 配置路由器接口

(1) AR1 的配置。

```
<Huawei>system-view
[Huawei]sysname AR1
[AR1]interface GigabitEthernet 0/0/0
[AR1-GigabitEthernet0/0/0] ip address 192.168.100.200 255.255.255.0
[AR1]interface GigabitEthernet 0/0/1
[AR1-GigabitEthernet0/0/1] ip address 192.168.101.1 255.255.255.0
```

(2) AR2 的配置。

```
[Huawei]system-view
[Huawei]sysname AR2
[AR2]interface GigabitEthernet 0/0/0
[AR2-GigabitEthernet0/0/0] ip address 192.168.100.201 255.255.255.0
[AR2]interface GigabitEthernet 0/0/1
[AR2-GigabitEthernet0/0/2] ip address 192.168.102.1 255.255.255.0
```

(3) AR3 的配置。

```
[Huawei]system-view
[Huawei]sysname AR3
[AR3]interface GigabitEthernet 0/0/1
[AR3-GigabitEthernet0/0/1]ip address 192.168.101.2 255.255.255.0
[AR3]interface GigabitEthernet 0/0/2
[AR3-GigabitEthernet0/0/2]ip address 192.168.103.1 255.255.255.0
```

(4) AR4 的配置。

```
[Huawei]system-view
[Huawei]sysname AR4
[AR4]interface GigabitEthernet 0/0/0
[AR4-GigabitEthernet0/0/1]ip address 192.168.102.2 255.255.255.0
[AR4]interface GigabitEthernet 0/0/2
[AR4-GigabitEthernet0/0/2] ip address 192.168.104.1 255.255.255.0
```

(5) AR5 的配置。

```
<Huawei>system-view
[Huawei]sysname AR5
[AR5]interface GigabitEthernet 0/0/0
[AR5-GigabitEthernet0/0/0]ip address 192.168.103.2 255.255.255.0
```

```
[AR5]interface GigabitEthernet 0/0/1
[AR5-GigabitEthernet0/0/1]ip address 192.168.104.2 255.255.255.0
[AR5]interface GigabitEthernet 0/0/2
[AR5-GigabitEthernet0/0/2]ip address 192.168.105.1 255.255.255.0
```

步骤二 部署 OSPF 协议

在路由器上配置 OSPF 协议，使用进程号 1，且所有网段均通告进区域 0 中。

（1）AR1 的配置。

```
[AR1]ospf 1
[AR1-ospf-1]area 0
[AR1-ospf-1-area-0.0.0.0]network 192.168.100.0 0.0.0.255
[AR1-ospf-1-area-0.0.0.0]network 192.168.101.0 0.0.0.255
```

（2）AR2 的配置。

```
[AR2]ospf 1
[AR2-ospf-1]area 0
[AR2-ospf-1-area-0.0.0.0]network 192.168.100.0 0.0.0.255
[AR2-ospf-1-area-0.0.0.0]network 192.168.102.0 0.0.0.255
```

（3）AR3 的配置。

```
[AR3]ospf 1
[AR3-ospf-1]area 0
[AR3-ospf-1-area-0.0.0.0]network 192.168.101.0 0.0.0.255
[AR3-ospf-1-area-0.0.0.0]network 192.168.103.0 0.0.0.255
```

（4）AR4 的配置。

```
[AR4]ospf 1
[AR4-ospf-1]area 0
[AR4-ospf-1-area-0.0.0.0]network 192.168.102.0 0.0.0.255
[AR4-ospf-1-area-0.0.0.0]network 192.168.104.0 0.0.0.255
```

（5）AR5 的配置。

```
[AR5]ospf 1
[AR5-ospf-1]area 0
[AR5-ospf-1-area-0.0.0.0]network 192.168.104.0 0.0.0.255
[AR5-ospf-1-area-0.0.0.0]network 192.168.103.0 0.0.0.255
[AR5-ospf-1-area-0.0.0.0]network 192.168.105.0 0.0.0.255
```

步骤三 配置 VRRP 协议

在 AR1、AR2 上配置 VRRP 协议，使用 vrrp vrid 1 virtual-ip 命令创建 VRRP 备份

组，指定路由器处于同一个 VRRP 备份组内，VRRP 备份组号为 1，配置虚拟 IP 为 192.168.100.254。

（1）AR1 的配置。

```
[AR1]interface G0/0/0
[AR1-GigabitEthernet0/0/0]vrrp vrid 1 virtual-ip 192.168.100.254
```

（2）AR2 的配置。

```
[AR2]interface GigabitEthernet 0/0/0
[AR2-GigabitEthernet0/0/0]vrrp vrid 1 virtual-ip 192.168.100.254
```

（3）配置 AR1 的优先级为 110，AR2 的优先级保持默认 100 不变，这样使得 AR1 成为主路由器，AR2 为备份路由器。

```
[AR1-GigabitEthernet0/0/0]vrrp vrid 1 priority 110
```

步骤四 配置上行接口监视

在 AR1 上配置上行接口监视，监视上行接口 G0/0/1，当此接口断掉时，裁剪优先级 60，使优先级变为 50，小于 R2 的优先级 100。

```
[AR1-GigabitEthernet0/0/0]vrrp vrid 1 track interface GigabitEthernet 0/0/1 reduced 60
```

步骤五 配置各部门计算机的 IP 地址

各部门计算机的 IP 地址配置如图 6-2、图 6-3 所示。

图 6-2 PC1 IP 地址配置图

图 6-3 服务器 IP 地址配置图

任务验证

步骤一 验证路由器上 OSPF 邻居建立信息

(1) AR1 的配置。

```
[AR1]dis ospf peer brief

        OSPF Process 1 with Router ID 192.168.100.200
            Peer Statistic Information
-----------------------------------------------------------------
Area Id          Interface              Neighbor id        State
0.0.0.0          GigabitEthernet0/0/0   192.168.100.201    Full
0.0.0.0          GigabitEthernet0/0/1   192.168.101.2      Full
-----------------------------------------------------------------
```

(2) AR2 的配置。

```
[AR2]dis ospf peer brief

        OSPF Process 1 with Router ID 192.168.100.201
            Peer Statistic Information
-----------------------------------------------------------------
Area Id          Interface              Neighbor id        State
0.0.0.0          GigabitEthernet0/0/0   192.168.100.200    Full
0.0.0.0          GigabitEthernet0/0/1   192.168.104.1      Full
-----------------------------------------------------------------
```

（3）AR3 的配置。

```
<AR3>dis ospf peer brief

       OSPF Process 1 with Router ID 192.168.101.2
           Peer Statistic Information
 -----------------------------------------------------------------
 Area Id         Interface                Neighbor id        State
 0.0.0.0         GigabitEthernet0/0/1     192.168.100.200    Full
 0.0.0.0         GigabitEthernet0/0/2     192.168.104.2      Full
 -----------------------------------------------------------------
```

（4）AR4 的配置。

```
<AR4>dis ospf peer brief

       OSPF Process 1 with Router ID 192.168.104.1
           Peer Statistic Information
 -----------------------------------------------------------------
 Area Id         Interface                Neighbor id        State
 0.0.0.0         GigabitEthernet0/0/2     192.168.104.2      Full
 0.0.0.0         GigabitEthernet0/0/0     192.168.100.201    Full
 -----------------------------------------------------------------
```

（5）AR5 的配置。

```
<AR5>dis ospf peer brief

       OSPF Process 1 with Router ID 192.168.104.2
           Peer Statistic Information
 -----------------------------------------------------------------
 Area Id         Interface                Neighbor id        State
 0.0.0.0         GigabitEthernet0/0/1     192.168.104.1      Full
 0.0.0.0         GigabitEthernet0/0/0     192.168.101.2      Full
 -----------------------------------------------------------------
```

步骤二 验证路由器 AR1、AR2 的 VRRP 信息

（1）AR1 的配置。

```
<AR1>dis vrrp
  GigabitEthernet0/0/0 | Virtual Router 1
    State : Master
    Virtual IP : 192.168.100.254
```

```
    Master IP : 192.168.100.200
    PriorityRun : 110
    PriorityConfig : 110
    MasterPriority : 110
    Preempt : YES    Delay Time : 0 s
    TimerRun : 1 s
    TimerConfig : 1 s
    Auth type : NONE
    Virtual MAC : 0000-5e00-0101
    Check TTL : YES
    Config type : normal-vrrp
    Backup-forward : disabled
    Track IF : GigabitEthernet0/0/1   Priority reduced : 60
    IF state : UP
    Create time : 2022-09-20 18:02:05 UTC-08:00
    Last change time : 2022-09-20 18:02:09 UTC-08:00
```

(2) AR2 的配置。

```
<AR2>dis vrrp
  GigabitEthernet0/0/0 | Virtual Router 1
    State : Backup
    Virtual IP : 192.168.100.254
    Master IP : 192.168.100.200
    PriorityRun : 100
    PriorityConfig : 100
    MasterPriority : 110
    Preempt : YES    Delay Time : 0 s
    TimerRun : 1 s
    TimerConfig : 1 s
    Auth type : NONE
    Virtual MAC : 0000-5e00-0101
    Check TTL : YES
    Config type : normal-vrrp
    Backup-forward : disabled
    Create time : 2022-09-20 18:02:39 UTC-08:00
    Last change time : 2022-09-20 18:32:48 UTC-08:00
```

可以观察到，现在 AR1 的 VRRP 状态是主路由器，AR2 是备份路由器。两者都处在 VRRP 备份组中。

步骤三 测试 PC 访问服务器时的数据包转发路径

```
PC>tracert 192.168.105.10

traceroute to 192.168.105.10, 8 hops max
(ICMP), press Ctrl+C to stop
 1  192.168.100.200    31 ms   47 ms   47 ms
 2  192.168.101.2      47 ms   31 ms   47 ms
 3  192.168.103.2      62 ms   47 ms   47 ms
 4  192.168.105.10     31 ms   47 ms   47 ms
```

步骤四 验证 VRRP 主备切换

将 AR1 的 GE 0/0/1 接口关闭：

```
[AR1]interface GigabitEthernet 0/0/1
[AR1-GigabitEthernet0/0/1]shutdown
```

经过 3s 左右，使用 display vrrp 查看 AR1、AR2 的 VRRP 信息。

（1）AR1 的配置。

```
[AR1-GigabitEthernet0/0/1]dis vrrp
  GigabitEthernet0/0/0 | Virtual Router 1
    State : Backup
    Virtual IP : 192.168.100.254
    Master IP : 192.168.100.201
    PriorityRun : 50
    PriorityConfig : 110
    MasterPriority : 100
    Preempt : YES    Delay Time : 0 s
    TimerRun : 1 s
    TimerConfig : 1 s
    Auth type : NONE
    Virtual MAC : 0000-5e00-0101
    Check TTL : YES
    Config type : normal-vrrp
    Backup-forward : disabled
    Track IF : GigabitEthernet0/0/1   Priority reduced : 60
    IF state : DOWN
    Create time : 2022-09-20 18:02:05 UTC-08:00
    Last change time : 2022-09-20 18:54:16 UTC-08:00
```

（2）AR2 的配置。

```
[AR2]dis vrrp
```

```
GigabitEthernet0/0/0 | Virtual Router 1
    State : Master
    Virtual IP : 192.168.100.254
    Master IP : 192.168.100.201
    PriorityRun : 100
    PriorityConfig : 100
    MasterPriority : 100
    Preempt : YES   Delay Time : 0 s
    TimerRun : 1 s
    TimerConfig : 1 s
    Auth type : NONE
    Virtual MAC : 0000-5e00-0101
    Check TTL : YES
    Config type : normal-vrrp
    Backup-forward : disabled
    Create time : 2022-09-20 18:02:39 UTC-08:00
    Last change time : 2022-09-20 18:54:17 UTC-08:00
```

可以观察到，AR1 切换成备份路由器，AR2 为主路由器，且 AR1 的 VRRP 优先级被裁减掉 60，变成 50，小于路由器 AR2 的优先级 100。

测试 PC 访问服务器时的数据包转发路径：

```
PC>tracert 192.168.105.10

traceroute to 192.168.105.10, 8 hops max
(ICMP), press Ctrl+C to stop
 1  192.168.100.201    47 ms   47 ms   47 ms
 2  192.168.102.2      63 ms   46 ms   47 ms
 3  192.168.104.2      32 ms   78 ms   62 ms
 4  192.168.105.10     31 ms   47 ms   47 ms
```

发现数据包发送路径已经切换。

任务 2　基于 VRRP 的负载均衡出口链路配置

任务背景

某公司采用 ISP-A、ISP-B 作为互联网接入服务商，通过出口路由器 AR1、AR2 连接，通过 VRRP 功能实现了路由器的主备自动切换。由于公司业务的开展，原来的主备链路模式无法满足出口带宽的需求,现需更改为负载均衡模式,在出口链路互为备份的同时还能分流出口流量,增加出口带宽。公司网络拓扑图如图 6-4 所示,项目具体要求如下:

(1) 路由器 AR1 和 AR2 通过拨号方式接入 ISP 网络 AR3 和 AR4,AR5 是互联网的一台路由器。

(2) 公司要求配置部分计算机从 AR1 访问 Internet,部分计算机从 AR2 访问 Internet,且要求当 AR1 或 AR2 链路故障时,自动切换到无故障链路上。这样既能保障充分利用出口流量,又能在出现链路故障时确保网络的连通性。

(3) 路由器间采用 OSPF 动态路由协议互联。

(4) 拓扑测试计算机和路由器的 IP 地址及接口信息如图 6-4 所示。

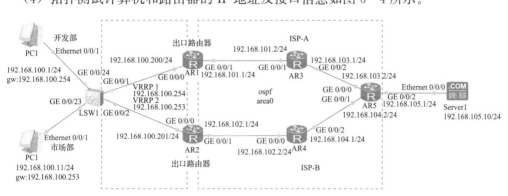

图 6-4　网络拓扑图

任务规划

为实现双出口的流量负载均衡,可以为不同的计算机指定不同的网关,使内部流量能通过不同的出口路由器进行转发。网关可以使用多个 VRRP 组来实现,并通过设置 VRRP 组中不同路由器的优先级,可使不同的虚拟网关以不同的路由器作为主路由器,另一个作为备份路由器,即可实现流量的分流和链路的备份。

公司有开发部和市场部 2 个部门,IP 地址段分别为 192.168.100.1-10/24 和 192.168.100.11-20/24,可以为不同的部门指定不同的出口网关以实现均衡流量。在 AR1、AR2 上创建 VRRP 1 和 VRRP 2,并创建虚拟网关地址 192.168.100.254 和 192.168.100.253 分别用于开发部和市场部的默认网关。其中,VRRP 1 的主路由器为 AR1,即 AR1 的 VRRP 1 优先级为 110,AR2 的优先级不变;VRRP 2 的主路由器为 AR2,即 AR2 的 VRRP 2 优先级为 110,AR1 的优先级不变。这样即可实现流量的负载均衡。同时,配置对 AR1、AR2 路由器上联接口的链路状态跟踪,当上联接口的链路状态为 DOWN 时,VRRP 主路由器优先级下降至 50,此时备份路由器即可切换为主路由器,实现自动的主备链路切换。在互联网连接方面,由于 ISP-A、ISP-B 均采用 OSPF 协议,故所有路由器均配置 OSPF 协议,并设置为 Area0 区域,以实现路由器之间的通信。

基于 VRRP 的负载均衡出口链路配置——任务实施

配置步骤如下。
(1) 配置路由器接口；
(2) 部署 OSPF 协议；
(3) 配置 VRRP 协议；
(4) 配置上行接口监视；
(5) 配置各部门计算机的 IP 地址。
IP 地址规划见表 6-2。

表 6-2 IP 地址规划表

设备	接口	IP 地址
AR1	GE 0/0/0	192.168.100.200/24
AR1	GE 0/0/1	192.168.101.1/24
AR2	GE 0/0/0	192.168.100.201/24
AR2	GE 0/0/1	192.168.102.1/24
AR3	GE 0/0/1	192.168.101.2/24
AR3	GE 0/0/2	192.168.103.1/24
AR4	GE 0/0/0	192.168.102.2/24
AR4	GE 0/0/2	192.168.104.1/24
AR5	GE 0/0/0	192.168.103.2/24
AR5	GE 0/0/1	192.168.104.2/24
AR5	GE 0/0/2	192.168.105.1/24

任务实施

步骤一 配置路由器接口

(1) AR1 的配置。

```
<Huawei>system-view
[Huawei]sysname AR1
[AR1]interface GigabitEthernet 0/0/0
[AR1-GigabitEthernet0/0/0] ip address 192.168.100.200 255.255.255.0
[AR1]interface GigabitEthernet 0/0/1
[AR1-GigabitEthernet0/0/1] ip address 192.168.101.1 255.255.255.0
```

(2) AR2 的配置。

```
[Huawei]system-view
[Huawei]sysname AR2
```

[AR2]interface GigabitEthernet 0/0/0

[AR2-GigabitEthernet0/0/0] ip address 192.168.100.201 255.255.255.0

[AR2]interface GigabitEthernet 0/0/1

[AR2-GigabitEthernet0/0/2] ip address 192.168.102.1 255.255.255.0

（3）AR3 的配置。

[Huawei]system-view

[Huawei]sysname AR3

[AR3]interface GigabitEthernet 0/0/1

[AR3-GigabitEthernet0/0/1]ip address 192.168.101.2 255.255.255.0

[AR3]interface GigabitEthernet 0/0/2

[AR3-GigabitEthernet0/0/2]ip address 192.168.103.1 255.255.255.0

（4）AR4 的配置。

[Huawei]system-view

[Huawei]sysname AR4

[AR4]interface GigabitEthernet 0/0/0

[AR4-GigabitEthernet0/0/1]ip address 192.168.102.2 255.255.255.0

[AR4]interface GigabitEthernet 0/0/2

[AR4-GigabitEthernet0/0/2] ip address 192.168.104.1 255.255.255.0

（5）AR5 的配置。

<Huawei>system-view

<Huawei>system-view

[Huawei]sysname AR5

[AR5]interface GigabitEthernet 0/0/0

[AR5-GigabitEthernet0/0/0]ip address 192.168.103.2 255.255.255.0

[AR5]interface GigabitEthernet 0/0/1

[AR5-GigabitEthernet0/0/1]ip address 192.168.104.2 255.255.255.0

[AR5]interface GigabitEthernet 0/0/2

[AR5-GigabitEthernet0/0/2]ip address 192.168.105.1 255.255.255.0

步骤二 部署 OSPF 协议

在路由器上配置 OSPF 协议，使用进程号 1，且所有网段均通告进 Area 0 区域中。

（1）AR1 的配置。

[AR1]ospf 1

[AR1-ospf-1]area 0

[AR1-ospf-1-area-0.0.0.0]network 192.168.100.0 0.0.0.255

[AR1-ospf-1-area-0.0.0.0]network 192.168.101.0 0.0.0.255

(2) AR2 的配置。

```
[AR2]ospf 1
[AR2-ospf-1]area 0
[AR2-ospf-1-area-0.0.0.0]network 192.168.100.0 0.0.0.255
[AR2-ospf-1-area-0.0.0.0]network 192.168.102.0 0.0.0.255
```

(3) AR3 的配置。

```
[AR3]ospf 1
[AR3-ospf-1]area 0
[AR3-ospf-1-area-0.0.0.0]network 192.168.101.0 0.0.0.255
[AR3-ospf-1-area-0.0.0.0]network 192.168.103.0 0.0.0.255
```

(4) AR4 的配置。

```
[AR4]ospf 1
[AR4-ospf-1]area 0
[AR4-ospf-1-area-0.0.0.0]network 192.168.102.0 0.0.0.255
[AR4-ospf-1-area-0.0.0.0]network 192.168.104.0 0.0.0.255
```

(5) AR5 的配置。

```
[AR5]ospf 1
[AR5-ospf-1]area 0
[AR5-ospf-1-area-0.0.0.0]network 192.168.104.0 0.0.0.255
[AR5-ospf-1-area-0.0.0.0]network 192.168.103.0 0.0.0.255
[AR5-ospf-1-area-0.0.0.0]network 192.168.105.0 0.0.0.255
```

步骤三 配置 VRRP 协议

在 AR1、AR2 上配置 VRRP 协议，使用 vrrp vrid 1 virtual-ip 命令创建 VRRP 备份组，指定路由器处于同一个 VRRP 备份组内。VRRP 备份组号为 1，配置虚拟 IP 为 192.168.100.254；VRRP 备份组号为 2，配置虚拟 IP 为 192.168.100.253。

(1) AR1 的配置。

```
[AR1]interface G0/0/0
[AR1-GigabitEthernet0/0/0]vrrp vrid 1 virtual-ip 192.168.100.254
[AR1-GigabitEthernet0/0/0]vrrp vrid 2 virtual-ip 192.168.100.253
```

(2) AR2 的配置。

```
[AR2]interface GigabitEthernet 0/0/0
[AR2-GigabitEthernet0/0/0]vrrp vrid 1 virtual-ip 192.168.100.254
[AR2-GigabitEthernet0/0/0]vrrp vrid 2 virtual-ip 192.168.100.253
```

配置 AR1 的备份组号 1 的 VRRP 优先级为 110，AR2 的备份组号 2 的 VRRP 优先级为 110。

(1) AR1 的配置。

```
[AR1-GigabitEthernet0/0/0]vrrp vrid 1 priority 110
```

(2) AR2 的配置。

```
[AR2-GigabitEthernet0/0/0]vrrp vrid 2 priority 110
```

步骤四 配置上行接口监视

在 AR1 上配置 VRRP 备份组 1 上行接口监视，监视上行接口 GE 0/0/1，当此接口断掉时，裁剪优先级 60，使优先级变为 50，小于 AR2 里 VRRP 备份组 1 的优先级 100。

```
[AR1-GigabitEthernet0/0/0]vrrp vrid 1 track interface GigabitEthernet 0/0/1 reduced 60
```

在 AR2 上配置 VRRP 备份组 2 上行接口监视，监视上行接口 GE 0/0/1，当此接口断掉时，裁剪优先级 60，使优先级变为 50，小于 AR1 里 VRRP 备份组 2 的优先级 100。

```
[AR2-GigabitEthernet0/0/0]vrrp vrid 2 track interface GigabitEthernet 0/0/1 reduced 60
```

步骤五 配置各部门计算机的 IP 地址

各部门计算机的 IP 地址配置如图 6-5、图 6-6、图 6-7 所示。

图 6-5 开发部 IP 地址配置图

图 6-6 市场部 IP 地址配置图

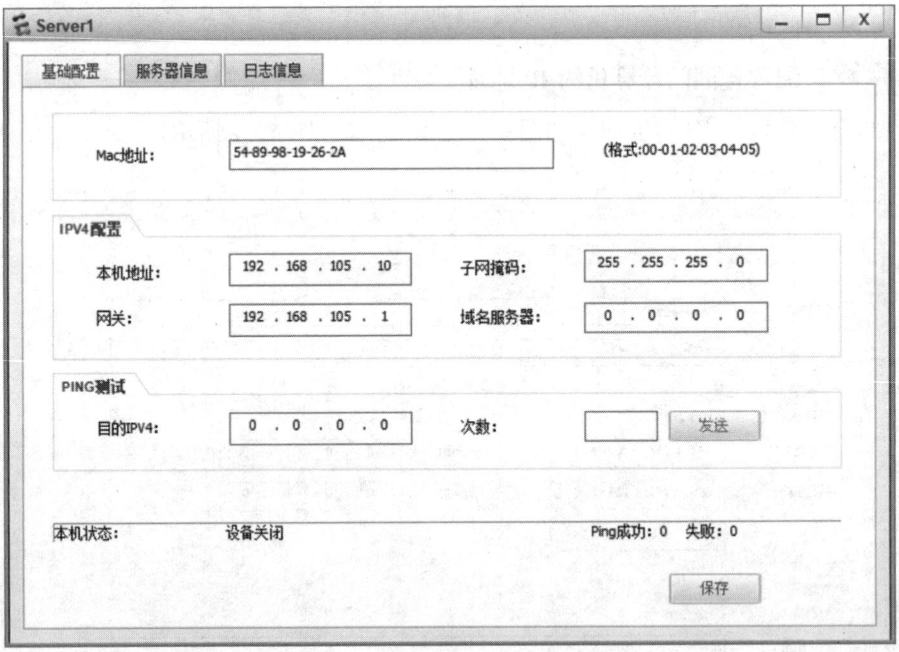

图 6-7 服务器 IP 地址配置图

任务验证

步骤一 验证路由器上 OSPF 邻居建立信息

(1) AR1 的配置。

```
[AR1]dis ospf peer brief

    OSPF Process 1 with Router ID 192.168.100.200
      Peer Statistic Information
 -----------------------------------------------------------------
 Area Id         Interface               Neighbor id         State
 0.0.0.0         GigabitEthernet0/0/0    192.168.100.201     Full
 0.0.0.0         GigabitEthernet0/0/1    192.168.101.2       FuLL
 -----------------------------------------------------------------
```

（2）AR2 的配置。

```
[AR2]dis ospf peer brief

    OSPF Process 1 with Router ID 192.168.100.201
      Peer Statistic Information
 -----------------------------------------------------------------
 Area Id         Interface               Neighbor id         State
 0.0.0.0         GigabitEthernet0/0/0    192.168.100.200     Full
 0.0.0.0         GigabitEthernet0/0/1    192.168.102.2       Full
 -----------------------------------------------------------------
```

（3）AR3 的配置。

```
<AR3>dis ospf peer brief

    OSPF Process 1 with Router ID 192.168.101.2
      Peer Statistic Information
 -----------------------------------------------------------------
 Area Id         Interface               Neighbor id         State
 0.0.0.0         GigabitEthernet0/0/1    192.168.100.200     Full
 0.0.0.0         GigabitEthernet0/0/2    192.168.104.2       Full
 -----------------------------------------------------------------
```

（4）AR4 的配置。

```
<AR4>dis ospf peer brief

    OSPF Process 1 with Router ID 192.168.102.2
      Peer Statistic Information
 -----------------------------------------------------------------
 Area Id         Interface               Neighbor id         State
 0.0.0.0         GigabitEthernet0/0/1    192.168.100.201     Full
 0.0.0.0         GigabitEthernet0/0/2    192.168.104.2       Full
 -----------------------------------------------------------------
```

(5) AR5 的配置。

```
<AR5>dis ospf peer brief


        OSPF Process 1 with Router ID 192.168.104.2
                Peer Statistic Information
 ---------------------------------------------------------------------
 Area Id            Interface                 Neighbor id         State
 0.0.0.0            GigabitEthernet0/0/1      192.168.102.2       Full
 0.0.0.0            GigabitEthernet0/0/0      192.168.101.2       Full
 ---------------------------------------------------------------------
```

步骤二 验证路由器 AR1、AR2 的 VRRP 信息

(1) AR1 的配置。

```
[AR1]dis vrrp
  GigabitEthernet0/0/0 | Virtual Router 1
    State : Master
    Virtual IP : 192.168.100.254
    Master IP : 192.168.100.200
    PriorityRun : 110
    PriorityConfig : 110
    MasterPriority : 110
    Preempt : YES   Delay Time : 0 s
    TimerRun : 1 s
    TimerConfig : 1 s
    Auth type : NONE
    Virtual MAC : 0000-5e00-0101
    Check TTL : YES
    Config type : normal-vrrp
    Backup-forward : disabled
    Track IF : GigabitEthernet0/0/1   Priority reduced : 60
    IF state : UP
    Create time : 2022-09-20 18:02:05 UTC-08:00
    Last change time : 2022-09-20 19:24:08 UTC-08:00

  GigabitEthernet0/0/0 | Virtual Router 2
    State : Backup
    Virtual IP : 192.168.100.253
    Master IP : 192.168.100.201
    PriorityRun : 100
    PriorityConfig : 100
```

　　　　MasterPriority：110
　　　　Preempt：YES　Delay Time：0 s
　　　　TimerRun：1 s
　　　　TimerConfig：1 s
　　　　Auth type：NONE
　　　　Virtual MAC：0000-5e00-0102
　　　　Check TTL：YES
　　　　Config type：normal-vrrp
　　　　Backup-forward：disabled
　　　　Create time：2022-09-20 19:18:24 UTC-08:00
　　　　Last change time：2022-09-20 19:24:09 UTC-08:00

（2）AR2 的配置。

　　[AR2]dis vrrp
　　　GigabitEthernet0/0/0 | Virtual Router 1
　　　　State：Backup
　　　　Virtual IP：192.168.100.254
　　　　Master IP：192.168.100.200
　　　　PriorityRun：100
　　　　PriorityConfig：100
　　　　MasterPriority：110
　　　　Preempt：YES　Delay Time：0 s
　　　　TimerRun：1 s
　　　　TimerConfig：1 s
　　　　Auth type：NONE
　　　　Virtual MAC：0000-5e00-0101
　　　　Check TTL：YES
　　　　Config type：normal-vrrp
　　　　Backup-forward：disabled
　　　　Create time：2022-09-20 19:32:14 UTC-08:00
　　　　Last change time：2022-09-20 19:32:17 UTC-08:00

　　　GigabitEthernet0/0/0 | Virtual Router 2
　　　　State：Master
　　　　Virtual IP：192.168.100.253
　　　　Master IP：192.168.100.201
　　　　PriorityRun：110
　　　　PriorityConfig：110
　　　　MasterPriority：110
　　　　Preempt：YES　Delay Time：0 s

```
    TimerRun : 1 s
    TimerConfig : 1 s
    Auth type : NONE
    Virtual MAC : 0000-5e00-0102
    Check TTL : YES
    Config type : normal-vrrp
    Backup-forward : disabled
    Track IF : GigabitEthernet0/0/1   Priority reduced : 60
    IF state : UP
    Create time : 2022-09-20 19:32:32 UTC-08:00
    Last change time : 2022-09-20 19:32:35 UTC-08:00
```

步骤三 测试 PC 访问服务器时的数据包转发路径

(1) 开发部。

```
PC>tracert 192.168.105.10

traceroute to 192.168.105.10, 8 hops max
(ICMP), press Ctrl+C to stop
 1  192.168.100.200    47 ms   46 ms   47 ms
 2  192.168.101.2      32 ms   46 ms   47 ms
 3  192.168.103.2      32 ms   62 ms   47 ms
 4  192.168.105.10     47 ms   47 ms   31 ms
```

(2) 市场部。

```
PC>tracert 192.168.105.10

traceroute to 192.168.105.10, 8 hops max
(ICMP), press Ctrl+C to stop
 1  192.168.100.201    37 ms   36 ms   47 ms
 2  192.168.102.2      22 ms   26 ms   47 ms
 3  192.168.104.2      52 ms   82 ms   47 ms
 4  192.168.105.10     27 ms   57 ms   31 ms
```

可以观察到,现在已经实现流量负载均衡。

步骤四 验证 VRRP 主备切换

将 AR1 的 GE 0/0/1 接口关闭:

```
[AR1]interface GigabitEthernet 0/0/1
[AR1-GigabitEthernet0/0/1]shutdown
```

经过 3s 左右,使用 display vrrp 查看 AR1、AR2 的 VRRP 信息。

（1）AR1 的配置。

```
[AR1]dis vrrp
  GigabitEthernet0/0/0 | Virtual Router 1
    State : Backup
    Virtual IP : 192.168.100.254
    Master IP : 192.168.100.201
    PriorityRun : 50
    PriorityConfig : 110
    MasterPriority : 100
    Preempt : YES    Delay Time : 0 s
    TimerRun : 1 s
    TimerConfig : 1 s
    Auth type : NONE
    Virtual MAC : 0000-5e00-0101
    Check TTL : YES
    Config type : normal-vrrp
    Backup-forward : disabled
    Track IF : GigabitEthernet0/0/1   Priority reduced : 60
    IF state : DOWN
    Create time : 2022-09-20 19:35:26 UTC-08:00
    Last change time : 2022-09-20 20:37:29 UTC-08:00

  GigabitEthernet0/0/0 | Virtual Router 2
    State : Backup
    Virtual IP : 192.168.100.253
    Master IP : 192.168.100.201
    PriorityRun : 100
    PriorityConfig : 100
    MasterPriority : 110
    Preempt : YES    Delay Time : 0 s
    TimerRun : 1 s
    TimerConfig : 1 s
    Auth type : NONE
    Virtual MAC : 0000-5e00-0102
    Check TTL : YES
    Config type : normal-vrrp
    Backup-forward : disabled
    Create time : 2022-09-20 19:38:35 UTC-08:00
    Last change time : 2022-09-20 20:40:23 UTC-08:00
```

（2）AR2 的配置。

```
<AR2>dis vrrp
  GigabitEthernet0/0/0 | Virtual Router 1
    State : Master
    Virtual IP : 192.168.100.254
    Master IP : 192.168.100.201
    PriorityRun : 100
    PriorityConfig : 100
    MasterPriority : 100
    Preempt : YES    Delay Time : 0 s
    TimerRun : 1 s
    TimerConfig : 1 s
    Auth type : NONE
    Virtual MAC : 0000-5e00-0101
    Check TTL : YES
    Config type : normal-vrrp
    Backup-forward : disabled
    Create time : 2022-09-20 19:43:45 UTC-08:00
    Last change time : 2022-09-20 20:55:38 UTC-08:00

  GigabitEthernet0/0/0 | Virtual Router 2
    State : Master
    Virtual IP : 192.168.100.253
    Master IP : 192.168.100.201
    PriorityRun : 110
    PriorityConfig : 110
    MasterPriority : 110
    Preempt : YES    Delay Time : 0 s
    TimerRun : 1 s
    TimerConfig : 1 s
    Auth type : NONE
    Virtual MAC : 0000-5e00-0102
    Check TTL : YES
    Config type : normal-vrrp
    Backup-forward : disabled
    Track IF : GigabitEthernet0/0/1   Priority reduced : 60
    IF state : UP
    Create time : 2022-09-20 19:45:55 UTC-08:00
    Last change time : 2022-09-20 20:55:23 UTC-08:00
```

可以观察到，AR1 切换成为备份路由器，AR2 切换为主路由器，且 AR1 的 VRRP 优先级被裁减掉 60，变成 50，小于路由器 AR2 的优先级 100。

测试 PC 访问服务器时的数据包转发路径：
（1）开发部。

```
PC>tracert 192.168.105.10

traceroute to 192.168.105.10, 8 hops max
(ICMP), press Ctrl+C to stop
 1  192.168.100.201    31 ms   47 ms   47 ms
 2  192.168.102.2      46 ms   32 ms   47 ms
 3  192.168.104.2      62 ms   63 ms   46 ms
 4  192.168.105.10     32 ms   47 ms   46 ms
```

（2）市场部。

```
PC>tracert 192.168.105.10

traceroute to 192.168.105.10, 8 hops max
(ICMP), press Ctrl+C to stop
 1  192.168.100.201    31 ms   47 ms   47 ms
 2  192.168.102.2      46 ms   32 ms   47 ms
 3  192.168.104.2      62 ms   63 ms   46 ms
 4  192.168.105.10     32 ms   47 ms   46 ms
```

此时我们发现数据包发送路径已经切换到 AR2。

任务 3　链路聚合

任务背景

某公司有市场部和行政部，各有计算机若干台，使用 2 台二层交换机进行局域网组建。为提高链路的传输带宽和冗余能力，交换机之间采用链路聚合的方式进行互联。同时，出于安全的考虑，两个部门的计算机划分在不同的 VLAN 中。网络拓扑图如图 6-8 所示。具体要求如下：

（1）两台交换机上均为市场部创建了 VLAN 5，为行政部创建了 VLAN 6，并将各部门 PC 连接交换机的接口划分到了对应的 VLAN；

（2）交换机 LSW1 通过 GE 0/0/3、GE 0/0/4 两个接口与 LSW2 互联，使用链路聚合提高带宽；

（3）计算机与交换机的接口信息如图 6-8 所示。

链路聚合

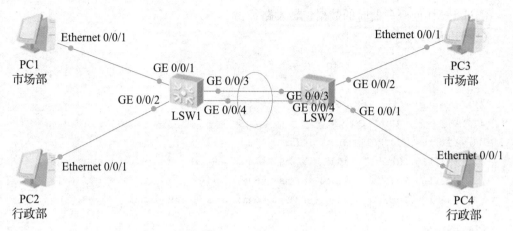

图 6-8 网络拓扑图

任务规划

2 台交换机使用 GE 0/0/3 和 GE 0/0/4 端口进行互联，采用链路聚合的方式提高传输带宽和冗余能力。创建 VLAN 5、VLAN 6 分别用于市场部和行政部，以实现两个部门的计算机相互隔离。因为两个部门的计算机连接在不同的交换机上，所以需要配置聚合链路为 Trunk 模式。

配置步骤如下：

（1）创建 VLAN；

（2）将端口划分至相应的 VLAN；

（3）创建 Eth-Trunk 1 接口，指定为手工负载分担模式；

（4）将接口加入 Eth-Trunk 1 接口；

（5）将链路划分至相应的 VLAN；

（6）配置各部门计算机的 IP 地址。

具体规划见表 6-3、表 6-4、表 6-5。

表 6-3 VLAN 规划表

VLAN ID	VLAN 命名	IP 地址段	用途
VLAN 5	MD	192.168.99.1-10/24	
VLAN 6	AD	192.168.99.11-20/24	

表 6-4 端口规划表

本端设备	端口号	端口类型	所属 VLAN	对端设备
LSW1	GE 0/0/1	access	VLAN 5	市场部 PC1
LSW1	GE 0/0/2	access	VLAN 6	行政部 PC2
LSW2	GE 0/0/2	access	VLAN 5	市场部 PC3
LSW2	GE 0/0/1	access	VLAN 6	行政部 PC4

表 6-5　IP 地址规划表

计算机	IP 地址
市场部 PC1	192.168.99.1/24
行政部 PC2	192.168.99.11/24
市场部 PC3	192.168.99.2/24
行政部 PC4	192.168.99.12/24

任务实施

步骤一　创建 VLAN

为各部门创建相应的 VLAN。

（1）LSW1 的配置。

```
[Huawei]system-view
[Huawei]sysname LSW1
[LSW1]vlan 5
[LSW1-vlan5]description MD
[LSW1]vlan 6
[LSW1-vlan6]description AD
```

（2）LSW2 的配置。

```
[Huawei]system-view
[Huawei]sysname LSW2
[LSW2]vlan 5
[LSW2-vlan5]description MD
[LSW2]vlan 6
[LSW2-vlan6]description AD
```

步骤二　将端口划分至相应的 VLAN

（1）LSW1 的配置。

```
[LSW1]interface GE0/0/1
[LSW1-GE0/0/1]port link-type access
[LSW1-GE0/0/1]port default vlan 5
[LSW1]interface GE0/0/2
[LSW1-GE0/0/2]port link-type access
[LSW1-GE0/0/2]port default vlan 6
```

(2) LSW2 的配置。

```
[LSW2]interface GE0/0/2
[LSW2-GE0/0/2]port link-type access
[LSW2-GE0/0/2]port default vlan 5
[LSW2]interface GE0/0/1
[LSW2-GE0/0/2]port link-type access
[LSW2-GE0/0/2]port default vlan 6
```

步骤三 创建 Eth-Trunk 1 接口，指定为手工负载分担模式

(1) LSW1 的配置。

```
[LSW1]interface Eth-Trunk 1
[LSW1-Eth-Trunk1]mode manual load-balance
```

(2) LSW2 的配置。

```
[LSW2]interface Eth-Trunk 1
[LSW2-Eth-Trunk1]mode manual load-balance
```

步骤四 将接口加入 Eth-Trunk 1 接口

将 LSW1 和 LSW2 的 GE 0/0/3 和 GE 0/0/4 分别加入 Eth-Trunk 1 接口。

(1) LSW1 的配置。

```
[LSW1]interface GigabitEthernet 0/0/3
[LSW1-GigabitEthernet0/0/3] eth-trunk 1
[LSW1]interface GigabitEthernet 0/0/4
[LSW1-GigabitEthernet0/0/4] eth-trunk 1
```

(2) LSW2 的配置。

```
[LSW2]interface GigabitEthernet 0/0/3
[LSW2-GigabitEthernet0/0/3] eth-trunk 1
[LSW2]interface GigabitEthernet 0/0/4
[LSW2-GigabitEthernet0/0/4] eth-trunk 1
```

步骤五 将链路划分至相应的 VLAN

(1) LSW1 的配置。

```
[LSW1]interface Eth-Trunk 1
[LSW1-Eth-Trunk1]port link-type trunk
[LSW1-Eth-Trunk1]port trunk allow-pass vlan 5 6
```

(2) LSW2 的配置。

```
[LSW2]interface Eth-Trunk 1
[LSW2-Eth-Trunk1]port link-type trunk
[LSW2-Eth-Trunk1]port trunk allow-pass vlan 5 6
```

步骤六 配置各部门计算机的 IP 地址

各部门计算机的 IP 地址配置如图 6-9～图 6-12 所示。

图 6-9　市场部-PC1 IP 地址配置图

图 6-10　行政部-PC2 IP 地址配置图

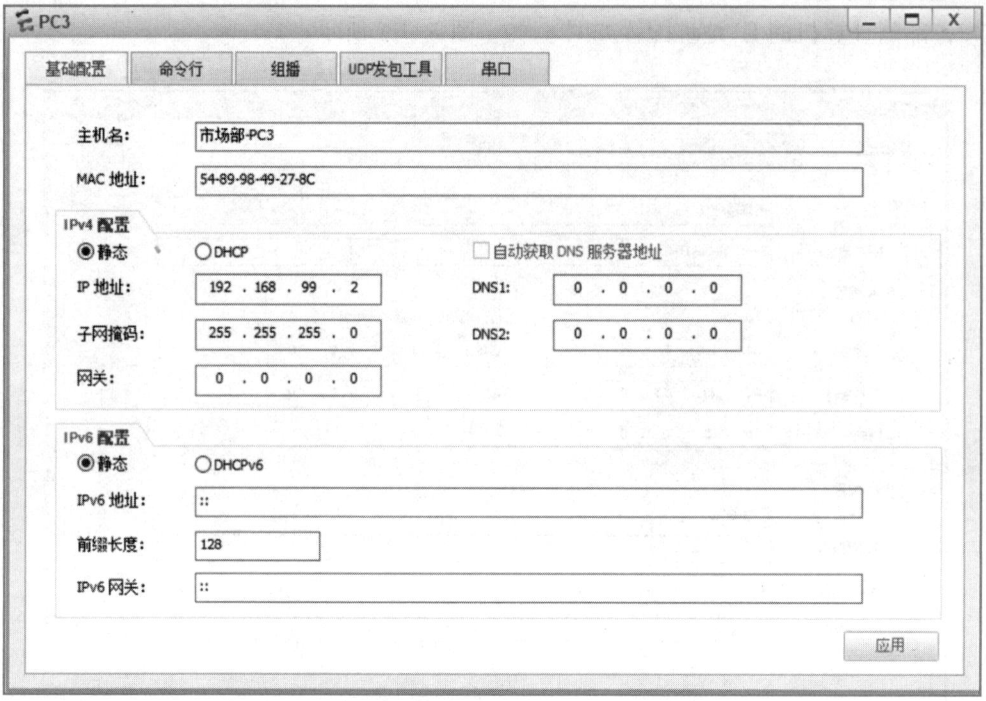

图 6-11　市场部-PC3 IP 地址配置图

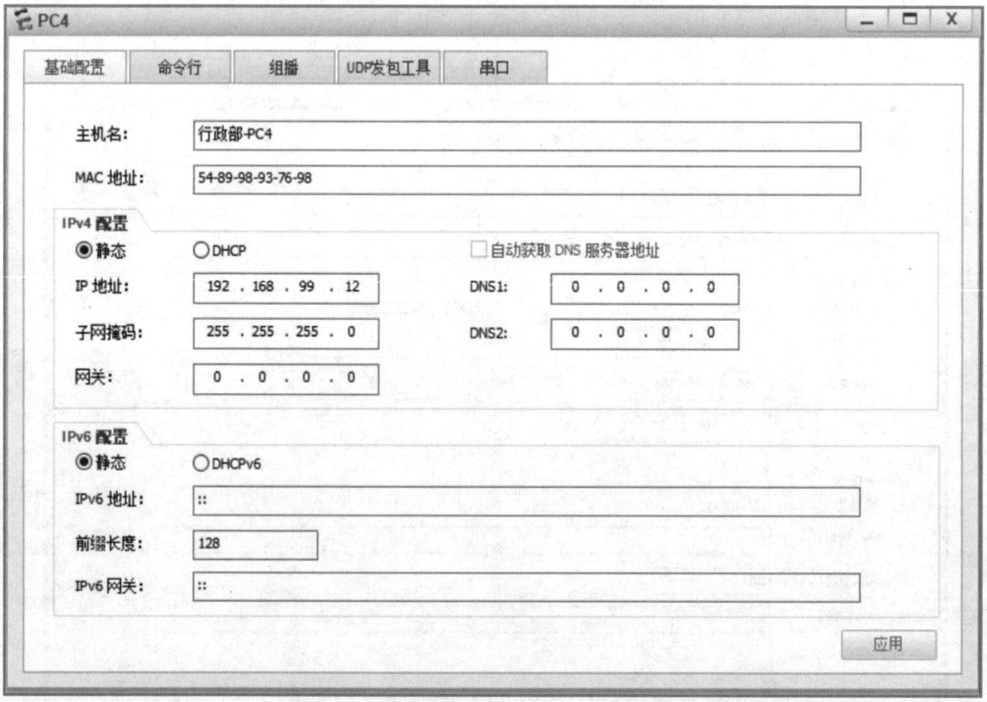

图 6-12　行政部-PC4 IP 地址配置图

任务验证

步骤一 验证 LSW1、LSW2 交换机的 VLAN 配置信息

（1）LSW1 的配置。

```
[LSW1]dis vlan
The total number of vlans is : 3
--------------------------------------------------------------------
U: Up;         D: Down;        TG: Tagged;       UT: Untagged;
MP: Vlan-mapping;              ST: Vlan-stacking;
#: ProtocolTransparent-vlan;   *: Management-vlan;
--------------------------------------------------------------------

VID  Type    Ports
--------------------------------------------------------------------
1    common  UT:GE0/0/5(D)    GE0/0/6(D)     GE0/0/7(D)     GE0/0/8(D)
             GE0/0/9(D)       GE0/0/10(D)    GE0/0/11(D)    GE0/0/12(D)
             GE0/0/13(D)      GE0/0/14(D)    GE0/0/15(D)    GE0/0/16(D)
             GE0/0/17(D)      GE0/0/18(D)    GE0/0/19(D)    GE0/0/20(D)
             GE0/0/21(D)      GE0/0/22(D)    GE0/0/23(D)    GE0/0/24(D)
             Eth-Trunk1(U)

5    common  UT:GE0/0/1(U)
             TG:Eth-Trunk1(U)
6    common  UT:GE0/0/2(U)
             TG:Eth-Trunk1(U)

VID  Status  Property    MAC-LRN Statistics Description
--------------------------------------------------------------------

1    enable  default     enable  disable    VLAN 0001
5    enable  default     enable  disable    MD
6    enable  default     enable  disable    AD
```

（2）LSW2 的配置。

```
[LSW2]dis vlan
The total number of vlans is : 3
--------------------------------------------------------------------
U: Up;         D: Down;        TG: Tagged;       UT: Untagged;
MP: Vlan-mapping;              ST: Vlan-stacking;
#: ProtocolTransparent-vlan;   *: Management-vlan;
--------------------------------------------------------------------
```

```
VID  Type    Ports
--------------------------------------------------------------------------
1    common  UT:GE0/0/5(D)    GE0/0/6(D)    GE0/0/7(D)    GE0/0/8(D)
                GE0/0/9(D)    GE0/0/10(D)   GE0/0/11(D)   GE0/0/12(D)
                GE0/0/13(D)   GE0/0/14(D)   GE0/0/15(D)   GE0/0/16(D)
                GE0/0/17(D)   GE0/0/18(D)   GE0/0/19(D)   GE0/0/20(D)
                GE0/0/21(D)   GE0/0/22(D)   GE0/0/23(D)   GE0/0/24(D)
                Eth-Trunk1(U)

5    common  UT:GE0/0/2(U)

             TG:Eth-Trunk1(U)

6    common  UT:GE0/0/1(U)

             TG:Eth-Trunk1(U)

VID  Status  Property    MAC-LRN Statistics Description
--------------------------------------------------------------------------
1    enable  default     enable  disable    VLAN 0001
5    enable  default     enable  disable    MD
6    enable  default     enable  disable    AD
```

步骤二 查看各交换机的 Eth-Trunk 1 接口状态

（1）LSW1 的配置。

```
[LSW1]display eth-trunk 1
Eth-Trunk1's state information is:
WorkingMode: NORMAL          Hash arithmetic: According to SIP-XOR-DIP
Least Active-linknumber: 1   Max Bandwidth-affected-linknumber: 8
Operate status: up           Number Of Up Port In Trunk: 2
--------------------------------------------------------------------------
PortName                     Status      Weight
GigabitEthernet0/0/3         Up          1
GigabitEthernet0/0/4         Up          1
```

（2）LSW2 的配置。

```
[LSW2]display eth-trunk 1
Eth-Trunk1's state information is:
WorkingMode: NORMAL          Hash arithmetic: According to SIP-XOR-DIP
```

```
Least Active-linknumber: 1   Max Bandwidth-affected-linknumber: 8
Operate status: up           Number Of Up Port In Trunk: 2
--------------------------------------------------------------
PortName                     Status        Weight
GigabitEthernet0/0/3         Up            1
GigabitEthernet0/0/4         Up            1
```

可以看到，该接口的总带宽是 GE 0/0/3 和 GE 0/0/4 接口的带宽之和。

步骤三 测试各部门计算机的互通性

通过 ping 命令，测试各部门内部通信情况。

使用市场部的计算机 ping 本部门的计算机：

```
PC>ping 192.168.99.2
Ping 192.168.99.2: 32 data bytes, Press Ctrl_C to break
From 192.168.99.2: bytes=32 seq=1 ttl=128 time=63 ms
From 192.168.99.2: bytes=32 seq=2 ttl=128 time=62 ms
From 192.168.99.2: bytes=32 seq=3 ttl=128 time=47 ms
From 192.168.99.2: bytes=32 seq=4 ttl=128 time=47 ms
From 192.168.99.2: bytes=32 seq=5 ttl=128 time=47 ms

---10.0.1.2 ping statistics ---
  5 packet(s) transmitted
  5 packet(s) received
  0.00%packet loss
  round-trip min/avg/max =47/53/63 ms
```

使用市场部的计算机 ping 行政部的计算机：

```
PC>ping 192.168.99.11
Ping 192.168.99.11: 32 data bytes, Press Ctrl_C to break
From 192.168.99.1: Destination host unreachable
From 192.168.99.1: Destination host unreachable
From 192.168.99.1: Destination host unreachable
From 192.168.99.1: Destination host unreachable
From 192.168.99.1: Destination host unreachable

---10.0.1.12 ping statistics ---
  5 packet(s) transmitted
  0 packet(s) received
  100.00%packet loss
```

项目小结

通过本项目的学习，同学们了解了 VRRP 和链路聚合技术的基本原理和应用场景。VRRP 通过部署多个网关解决单点故障问题，实现链路的负载均衡。链路聚合通过将多个物理接口捆绑为一个逻辑接口，增加链路带宽，同时采用备份链路的机制，提高设备间链路的可靠性。

项目 7 广域网技术

项目目标

1. 掌握 PAP 认证的配置方法
2. 掌握 CHAP 认证的配置方法
3. 掌握基于 PPPoE 认证的公司出口配置方法

任务 1 基于 PAP 认证的公司与分部安全互联

任务背景

某公司因业务发展需要建立了分公司,租用了专门的线路用于总部与分公司的互联。为保障通信线路的数据安全,该公司需要在路由器上配置安全认证。网络拓扑图如图 7-1 所示。具体要求如下:

(1) 公司总部路由器 R1 使用 S4/0/0 接口与分公司路由器 R2 互联;

(2) R1 的 S4/0/0 接口上使用 PPP 协议并启用 PAP 认证,用于分公司的安全接入;

(3) 计算机、路由器的 IP 地址及接口信息如图 7-1 所示。

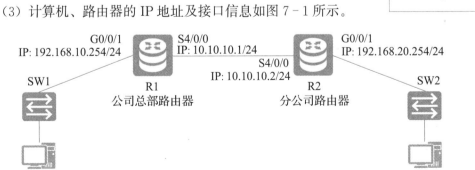

图 7-1 网络拓扑图

任务规划

串行链路默认采用 PPP 封装协议，可以通过 PAP 认证使链路的建立更安全。PAP 认证通过用户名和密码进行验证。公司总部路由器 R1 作为认证方，需在 AAA 视图下添加名为 Jan16 的 Local-user 用户，密码为 123456，并将接口 S4/0/0 的认证方式设置为 PAP；分公司路由器 R2 为被认证方，需在接口上配置 PAP 的认证方式，并添加与认证方一致的用户名和密码，即可实现链路的认证接入。

配置步骤如下：

（1）配置路由器接口；

（2）搭建 OSPF 网络；

（3）配置 PPP 的 PAP 认证；

（4）对端配置 PAP 认证；

（5）配置各计算机的 IP 地址。

具体规划见表 7-1、表 7-2。

表 7-1 IP 地址规划表

设备	接口	IP 地址
R1	G0/0/1	192.168.10.254/24
R1	S4/0/0	10.10.10.1/24
R2	G0/0/1	192.168.20.254/24
R2	S4/0/0	10.10.10.2/24
PC1	E0/0/1	192.168.10.1/24
PC2	E0/0/1	192.168.20.1/24

表 7-2 接口规划表

本端设备	接口	端口 IP 地址	对端设备
R1	G0/0/1	192.168.10.254	SW1
R1	S4/0/0	10.10.10.1/24	R2
R2	G0/0/1	192.168.20.254/24	SW2
R2	S4/0/0	10.10.10.2/24	R1

任务实施

步骤一 配置路由器接口

（1）R1 的配置。

```
[Huawei]system-view
[Huawei]sysname R1
[R1]int G0/0/1
[R1-GigabitEthernet0/0/1]ip add 192.168.10.254 24
[R1]interface Serial 4/0/0
[R1-Serial4/0/0]ip add 10.10.10.1 24
```

（2）R2 的配置。

```
[Huawei]system-view
[Huawei]sysname R2
[R1]int G0/0/1
[R1-GigabitEthernet0/0/1]ip add 192.168.20.254 24
[R1]interface Serial 4/0/0
[R1-Serial4/0/0]ip add 10.10.10.2 24
```

步骤二 搭建 OSPF 网络

（1）R1 的配置。

```
[R1]ospf 1
[R1-ospf-1]area 0
[R1-ospf-1-area-0.0.0.0]network 192.168.10.0 0.0.0.255
[R1-ospf-1-area-0.0.0.0]network 10.10.10.0 0.0.0.255
```

（2）R2 的配置。

```
[R2]ospf 1
[R2-ospf-1]area 0
[R2-ospf-1-area-0.0.0.0]network 192.168.20.0 0.0.0.255
[R2-ospf-1-area-0.0.0.0]network 10.10.10.0 0.0.0.255
```

步骤三 配置 PPP 的 PAP 认证

R1 路由器作为认证端，需要配置本端 PPP 协议的认证方式为 PAP。执行 aaa 命令，进入 AAA 视图，配置 PAP 认证所使用的用户名和密码。

```
[R1]aaa
[R1-aaa]local-user Jan16 password cipher 123456
[R1-aaa]local-user Jan16 service-type ppp
[R1-aaa]int s4/0/0
[R1-Serial4/0/0]link-protocol ppp
[R1-Serial4/0/0]ppp authentication-mode pap
```

配置完成后，关闭 R1 与 R2 相连接口一段时间后再打开，使 R1 与 R2 间的链路重新协商，并检查链路状态和连通性。

```
[R1]interface Serial 4/0/0
[R1-Serial4/0/0]shutdown
[R1-Serial4/0/0]undo shutdown

[R1]dis ip interface brief
*down: administratively down
^down: standby
```

```
  (l): loopback
  (s): spoofing
The number of interface that is UP in Physical is 3
The number of interface that is DOWN in Physical is 3
The number of interface that is UP in Protocol is 2
The number of interface that is DOWN in Protocol is 4

Interface                 IP Address/Mask        Physical    Protocol
GigabitEthernet0/0/0      unassigned             down        down
GigabitEthernet0/0/1      192.168.10.254/24      up          up
GigabitEthernet0/0/2      unassigned             down        down
NULL0                     unassigned             up          up(s)
Serial4/0/0               10.10.10.1/24          up          down
Serial4/0/1               unassigned             down        down

[R1]ping 10.10.10.2
  PING 10.10.10.2: 56   data bytes, press CTRL+C to break
    Request time out
    Request time out
    Request time out
    Request time out
    Request time out

  ---10.10.10.2 ping statistics ---
    5 packet(s) transmitted
    0 packet(s) received
    100.00% packet loss
```

可以观察到，现在 R1 与 R2 间无法正常通信，链路物理状态正常，但是链路层协议状态不正常。这是因为此时 PPP 链路上的 PAP 认证未通过。

步骤四 **对端配置 PAP 认证**

R2 作为被认证端，在 S4/0/0 接口下配置以 PAP 方式验证时本地发送的 PAP 用户名和密码。

```
[R2]int s4/0/0
[R2-Serial4/0/0]link-protocol ppp
[R2-Serial4/0/0]ppp pap local-user Jan16 password cipher 123456
```

步骤五 **配置各计算机的 IP 地址**

各计算机的 IP 地址配置如图 7-2、图 7-3 所示。

项目 7 广域网技术

图 7-2 PC1 IP 地址配置图

图 7-3 PC2 IP 地址配置图

路由交换技术

▊ 任务验证

步骤一 查看链路状态

R2 的配置：

```
[R2]dis ip int brief
*down: administratively down
^down: standby
(l): loopback
(s): spoofing
The number of interface that is UP in Physical is 3
The number of interface that is DOWN in Physical is 3
The number of interface that is UP in Protocol is 3
The number of interface that is DOWN in Protocol is 3

Interface                IP Address/Mask      Physical    Protocol
GigabitEthernet0/0/0     unassigned           down        down
GigabitEthernet0/0/1     192.168.20.254/24    up          up
GigabitEthernet0/0/2     unassigned           down        down
NULL0                    unassigned           up          up(s)
Serial4/0/0              10.10.10.2/24        up          up
Serial4/0/1              unassigned           down        down
```

可以观察到，现在 R1 与 R2 间的链路层协议状态正常。

步骤二 测试各计算机的互通性

使用 PC1 计算机 ping PC2 计算机：

```
PC>ping 192.168.20.1

Ping 192.168.20.1: 32 data bytes, Press Ctrl_C to break
From 192.168.20.1: bytes=32 seq=1 ttl=126 time=63 ms
From 192.168.20.1: bytes=32 seq=2 ttl=126 time=78 ms
From 192.168.20.1: bytes=32 seq=3 ttl=126 time=62 ms
From 192.168.20.1: bytes=32 seq=4 ttl=126 time=47 ms
From 192.168.20.1: bytes=32 seq=5 ttl=126 time=63 ms

---192.168.20.1 ping statistics ---
  5 packet(s) transmitted
  5 packet(s) received
  0.00%packet loss
  round-trip min/avg/max = 47/62/78 ms
```

可以观察到，计算机间正常通信。

任务 2　基于 CHAP 认证的公司与分部安全互联

任务背景

某公司因业务发展需要，建立了分公司，租用了专门的线路用于总部与分公司的互联。为保障通信线路的数据安全，该公司需要在路由器上配置安全认证。网络拓扑图如图 7-4 所示。具体要求如下：

（1）公司总部路由器 R1 使用 S4/0/0 接口与分公司路由器 R2 互联；

（2）R1 的 S4/0/0 接口上使用 PPP 协议并启用 CHAP 认证，用于分公司的安全接入；

（3）计算机和路由器的 IP 地址及接口信息如图 7-4 所示。

微课视频

基于 CHAP 认证的公司与分部安全互联

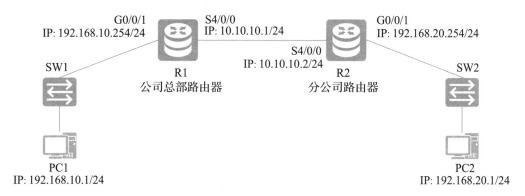

图 7-4　网络拓扑图

任务规划

串行链路默认采用 PPP 封装协议，可以通过 CHAP 认证使链路的建立更安全。CHAP 认证是由认证服务器向被认证方提出认证需求，通过用户名和密码进行验证。公司总部路由器 R1 作为认证方，需在 AAA 视图下添加名为 Jan16 的 Local-user 用户，密码为 123456，并将接口 S4/0/0 设置为 CHAP 的认证方式；分公司路由器 R2 为被认证方，需在接口上配置 CHAP 的认证方式，并添加与认证方一致的用户名和密码，即可实现链路的认证接入。

配置步骤如下：

（1）配置路由器接口；

（2）搭建 OSPF 网络；

（3）配置 PPP 的 CHAP 认证；

(4) 对端配置 CHAP 认证；
(5) 配置各计算机的 IP 地址。

具体规划见表 7-3、表 7-4。

表 7-3 IP 地址规划表

设备	接口	IP 地址
R1	G0/0/1	192.168.10.254/24
R1	S4/0/0	10.10.10.1/24
R2	G0/0/1	192.168.20.254/24
R2	S4/0/0	10.10.10.2/24
PC1	E0/0/1	192.168.10.1/24
PC2	E0/0/1	192.168.20.1/24

表 7-4 接口规划表

本端设备	接口	端口 IP 地址	对端设备
R1	G0/0/1	192.168.10.254	SW1
R1	S4/0/0	10.10.10.1/24	R2
R2	G0/0/1	192.168.20.254/24	SW2
R2	S4/0/0	10.10.10.2/24	R1

任务实施

步骤一 配置路由器接口

(1) R1 的配置。

```
[Huawei]system-view
[Huawei]sysname R1
[R1]int G0/0/1
[R1-GigabitEthernet0/0/1]ip add 192.168.10.254 24
[R1]interface Serial 4/0/0
[R1-Serial4/0/0]ip add 10.10.10.1 24
```

(2) R2 的配置。

```
[Huawei]system-view
[Huawei]sysname R2
[R1]int G0/0/1
[R1-GigabitEthernet0/0/1]ip add 192.168.20.254 24
```

```
[R1]interface Serial 4/0/0
[R1-Serial4/0/0]ip add 10.10.10.2 24
```

步骤二 搭建 OSPF 网络

（1）R1 的配置。

```
[R1]ospf 1
[R1-ospf-1]area 0
[R1-ospf-1-area-0.0.0.0]network 192.168.10.0 0.0.0.255
[R1-ospf-1-area-0.0.0.0]network 10.10.10.0 0.0.0.255
```

（2）R2 的配置。

```
[R2]ospf 1
[R2-ospf-1]area 0
[R2-ospf-1-area-0.0.0.0]network 192.168.20.0 0.0.0.255
[R2-ospf-1-area-0.0.0.0]network 10.10.10.0 0.0.0.255
```

步骤三 配置 PPP 的 CHAP 认证

R1 路由器作为认证端，需要配置本端 PPP 协议的认证方式为 CHAP。执行 aaa 命令，进入 AAA 视图，配置 CHAP 认证所使用的用户名和密码。

```
[R1]aaa
[R1-aaa]local-user Jan16 password cipher 123456
[R1-aaa]local-user Jan16 service-type ppp
[R1]interface Serial 4/0/0
[R1-Serial4/0/0]link-protocol ppp
[R1-Serial4/0/0]ppp authentication-mode chap
```

配置完成后，关闭 R1 与 R2 相连接口一段时间后再打开，使 R1 与 R2 间的链路重新协商，并检查链路状态和连通性。

```
[R1]interface Serial 4/0/0
[R1-Serial4/0/0]shutdown
[R1-Serial4/0/0]undo shutdown

[R1]display ip interface brief
*down: administratively down
^down: standby
(l): loopback
(s): spoofing
The number of interface that is UP in Physical is 3
The number of interface that is DOWN in Physical is 3
```

```
The number of interface that is UP in Protocol is 2
The number of interface that is DOWN in Protocol is 4


Interface                IP Address/Mask      Physical    Protocol
GigabitEthernet0/0/0     unassigned           down        down
GigabitEthernet0/0/1     192.168.10.254/24    up          up
GigabitEthernet0/0/2     unassigned           down        down
NULL0                    unassigned           up          up(s)
Serial4/0/0              10.10.10.1/24        up          down
Serial4/0/1              unassigned           down        down

[R1]ping 10.10.10.2
  PING 10.10.10.2: 56   data bytes, press CTRL_C to break
    Request time out
    Request time out
    Request time out
    Request time out
    Request time out

  ---10.10.10.2 ping statistics ---
    5 packet(s) transmitted
    0 packet(s) received
    100.00%packet loss
```

可以观察到，现在 R1 与 R2 间无法正常通信，链路的物理状态正常，但是链路层协议状态不正常。这是因为此时 PPP 链路上的 CHAP 认证未通过。

步骤四 对端配置 CHAP 认证

R2 作为被认证端，在 S4/0/0 接口下配置以 PAP 方式验证时本地发送的 CHAP 用户名和密码。

```
[R2]int s4/0/0
[R2-Serial4/0/0]link-protocol ppp
[R2-Serial4/0/0]ppp chap user Jan16
[R2-Serial4/0/0]ppp chap password 123456
```

步骤五 配置各计算机的 IP 地址

各计算机的 IP 地址配置如图 7-5、图 7-6 所示。

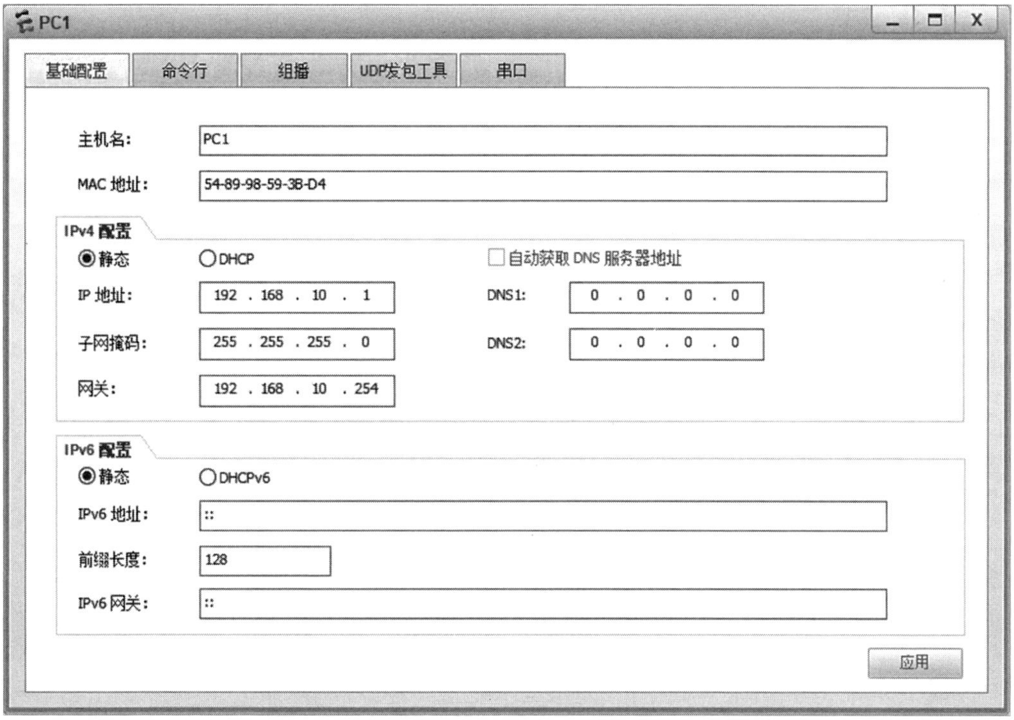

图 7-5　PC1 IP 地址配置图

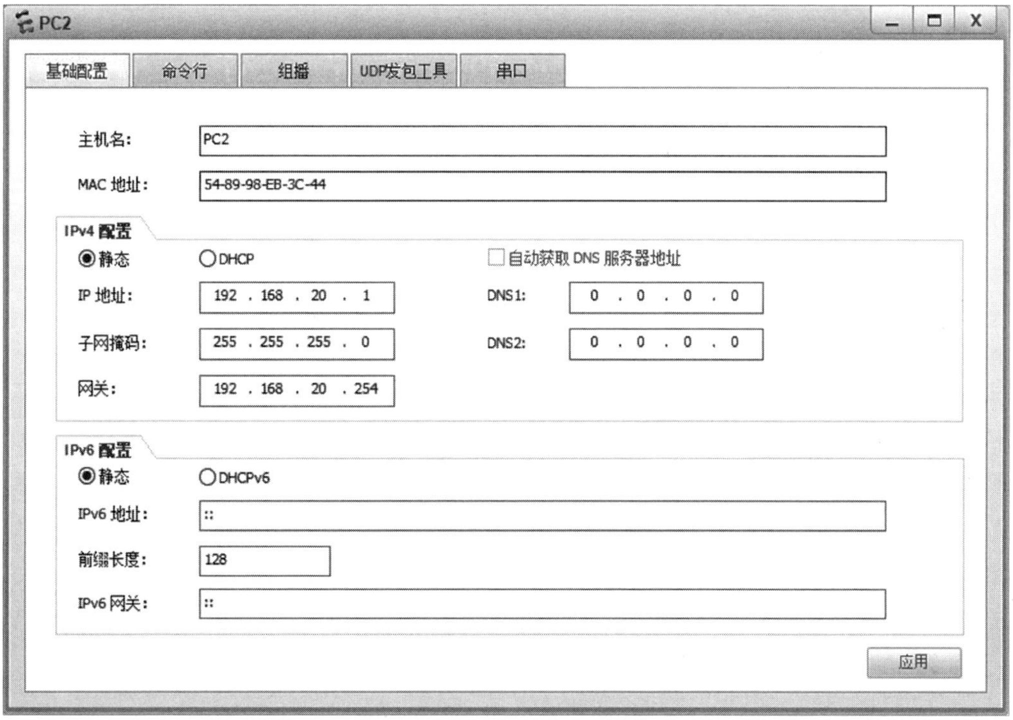

图 7-6　PC2 IP 地址配置图

任务验证

步骤一 查看链路状态

R2 的配置：

```
[R2]display ip interface brief
*down: administratively down
^down: standby
(l): loopback
(s): spoofing
The number of interface that is UP in Physical is 3
The number of interface that is DOWN in Physical is 3
The number of interface that is UP in Protocol is 3
The number of interface that is DOWN in Protocol is 3

Interface                 IP Address/Mask       Physical    Protocol
GigabitEthernet0/0/0      unassigned            down        down
GigabitEthernet0/0/1      192.168.20.254/24     up          up
GigabitEthernet0/0/2      unassigned            down        down
NULL0                     unassigned            up          up(s)
Serial4/0/0               10.10.10.2/24         up          up
Serial4/0/1               unassigned            down        down
```

可以观察到，现在 R1 与 R2 间的链路层协议状态正常。

步骤二 测试各计算机的互通性

使用 PC1 计算机 ping PC2 计算机：

```
PC>ping 192.168.20.1

Ping 192.168.20.1: 32 data bytes, Press Ctrl_C to break
From 192.168.20.1: bytes=32 seq=1 ttl=126 time=63 ms
From 192.168.20.1: bytes=32 seq=2 ttl=126 time=62 ms
From 192.168.20.1: bytes=32 seq=3 ttl=126 time=47 ms
From 192.168.20.1: bytes=32 seq=4 ttl=126 time=78 ms
From 192.168.20.1: bytes=32 seq=5 ttl=126 time=78 ms

---192.168.20.1 ping statistics ---
  5 packet(s) transmitted
  5 packet(s) received
  0.00%packet loss
  round-trip min/avg/max =47/65/78 ms
```

可以观察到，PC 间通信正常。

任务 3　基于 PPPoE 认证的公司出口配置

任务背景

某公司向 ISP 服务提供商申请了一条专用线路用于互联网接入，线路采用 PPPoE 接入方式。现需配置出口路由器，使内网可以通过共享出口路由器访问互联网。网络拓扑图如图 7-7 所示。具体要求如下：

（1）公司出口路由器采用 G0/0/0 接口与 ISP 认证服务器互联；

（2）公司内网通过一台无配置的交换机互联，实现客户端与网关的通信；

基于 PPPoE 认证的公司出口配置

（3）计算机和路由器的 IP 地址及接口信息如图 7-7 所示。

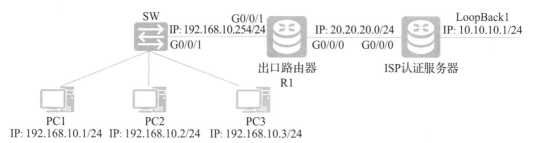

图 7-7　网络拓扑图

任务规划

PPPoE 接入采用 PPP 封装协议，可选择 PAP 或 CHAP 的认证方式，这里选择较为安全的 CHAP 认证。首先，建立 PPPoE 接入，需创建 Dialer 接口，并配置相应的认证方式、用户名和密码；其次，为实现接口共享，需通过 ACL 列表匹配内网流量并绑定到 Dialer 接口；最后，将 Dialer 接口与实际连接 ISP 服务器的 G0/0/0 接口进行绑定，即可实现链路的认证接入。为实现内网到互联网的访问，还需要在 R1 上配置默认路由，并指向 Dialer 接口。

配置步骤如下：

（1）配置 PPPoE 服务器；

（2）配置 PPPoE 客户端；

（3）配置各计算机的 IP 地址。

具体规划见表 7-5、表 7-6。

表 7-5 IP 地址规划表

设备	接口	IP 地址
ISP	G0/0/0	
ISP	LoopBack1	10.10.10.1/24
PC1	E0/0/1	192.168.10.1/24
PC2	E0/0/1	192.168.10.2/24
PC3	E0/0/1	192.168.10.3/24

表 7-6 接口规划表

本端设备	接口	端口 IP 地址	对端设备
R1	G0/0/1	192.168.10.254	SW1
R1	G0/0/0		ISP
ISP	G0/0/0	20.20.20.1/24	R1

任务实施

步骤一 配置 PPPoE 服务器

(1) 修改 ISP 路由器的设备名，并配置 LoopBack1 接口的 IP。

```
[Huawei]system-view
[Huawei]sysname ISP
[ISP] interface LoopBack1
[ISP-LoopBack1] ip address 10.10.10.1 255.255.255.0
```

(2) 配置 PPPoE 地址池，通过使用全局地址池给对端分配地址，实现 PPPoE Server 为 PPPoE Client 动态分配 IP 地址。

```
[ISP]ip pool pppoe
[ISP-ip-pool-pppoe]gateway-list 20.20.20.1
[ISP-ip-pool-pppoe]network 20.20.20.0 mask 255.255.255.0
```

(3) 配置 PPPoE 认证用户，实现 PPPoE Server 对用户主机的认证。

```
[ISP]aaa
[ISP-aaa]local-user r1 password cipher 123456
[ISP-aaa]local-user r1 privilege level 0
[ISP-aaa]local-user r1 service-type ppp
```

(4) 配置虚拟接口模板 VT，本端 PPPoE 协议对对端设备的认证方式为 CHAP。

```
[ISP]interface Virtual-Template 1
[ISP-Virtual-Template1]ppp authentication-mode chap
[ISP-Virtual-Template1]remote address pool pppoe
[ISP-Virtual-Template1]ip address 20.20.20.1 255.255.255.0
```

(5) 启用 PPPoE Server 功能,在以太网接口 G0/0/0 上启用 PPPoE Server 功能。

```
[ISP]interface GigabitEthernet 0/0/0
[ISP-GigabitEthernet0/0/0]pppoe-server bind Virtual-Template 1
```

步骤二 配置 PPPoE 客户端

(1) 修改 R1 路由器的设备名,并配置 G0/0/1 接口的 IP 作为内网用户的网关。

```
[Huawei]system-view
[Huawei]sysname R1
[R1]int G0/0/1
[R1-GigabitEthernet0/0/1] ip address 192.168.10.254 255.255.255.0
```

(2) 配置 Dialer 接口。

```
[R1]interface Dialer0
[R1-Dialer0]ppp chap user r1
[R1-Dialer0]ppp chap password cipher 123456
[R1-Dialer0]tcp adjust-mss 1200
[R1-Dialer0]ip address ppp-negotiate
[R1-Dialer0]dialer user isp
[R1-Dialer0]dialer bundle 1
[R1-Dialer0]dialer-group 1
```

(3) 建立 PPPoE 会话。

```
[R1]interface GigabitEthernet0/0/0
[R1-GigabitEthernet0/0/0] pppoe-client dial-bundle-number 1
```

(4) 配置 NAT 转换,配置局域网用户通过 NAT 转换将私网地址转换为公网地址,进行拨号上网。

```
[R1]acl number 3000
[R1-acl-adv-3000]rule 5 permit ip source 192.168.10.0 0.0.0.255
[R1-acl-adv-3000]quit
[R1]interface Dialer0
[R1-Dialer0]nat outbound 3000
```

(5) 配置到 PPPoE Server 的静态路由。

```
[R1] ip route-static 0.0.0.0 0 dialer 1
```

步骤三 配置各计算机的 IP 地址

各计算机的 IP 地址配置如图 7-8～图 7-10 所示。

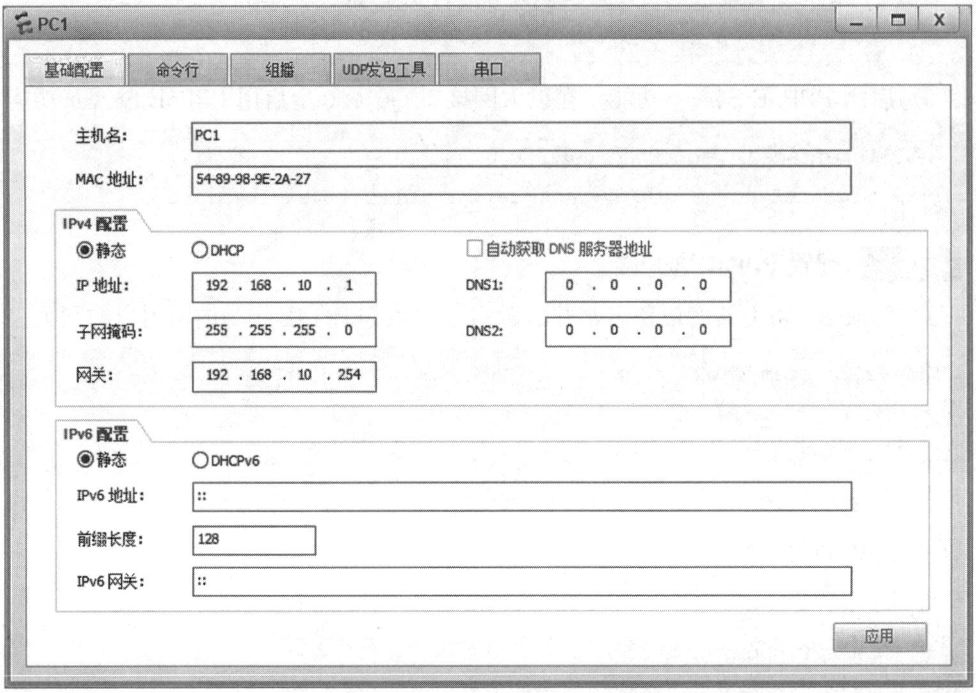

图 7-8 PC1 IP 地址配置图

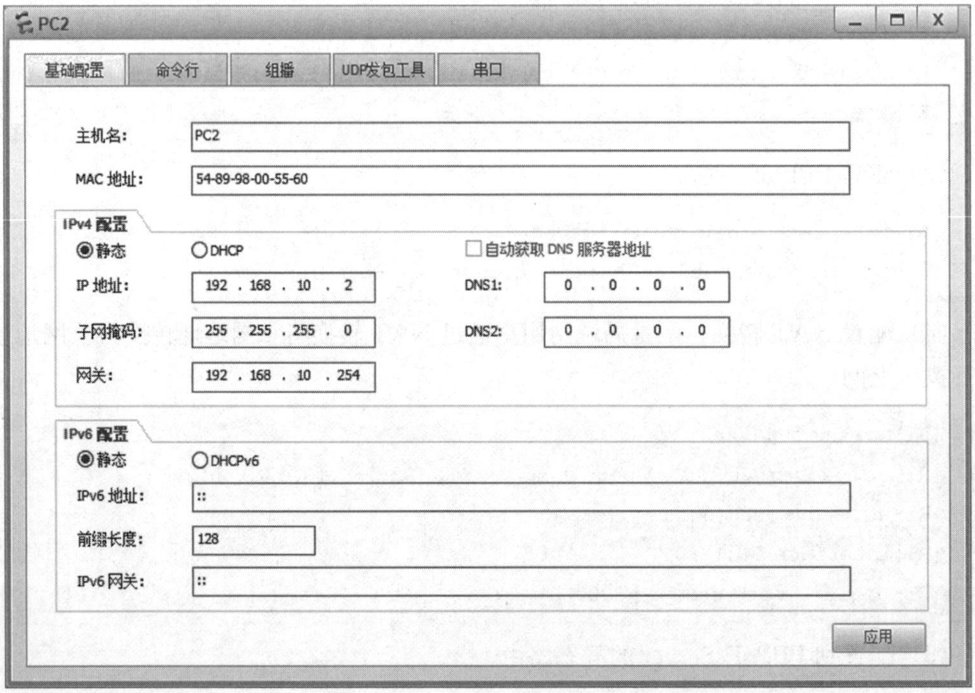

图 7-9 PC2 IP 地址配置图

图 7 - 10　PC3 IP 地址配置图

任务验证

步骤一　查看 PPPoE Server 会话的状态和配置信息

在 ISP 上使用 display pppoe-server session all 命令查看 pppoe 会话信息。

```
<ISP>display pppoe-server session all
SID  Intf              State  OIntf    RemMAC           LocMAC
1    Virtual-Template1:0  UP   GE0/0/0  00e0.fcdc.5981  00e0.fc96.67e0
```

可以观察到，会话状态正常（状态为 UP 表示正常）、配置正确（与之前的数据规划和组网一致）。

步骤二　查看 PPPoE Client 会话状态和配置信息

在 R1 上使用 display pppoe-client session summary 命令查看 pppoe 会话信息。

```
[R1]display pppoe-client session summary
PPPoE Client Session:
ID  Bundle  Dialer  Intf      Client-MAC    Server-MAC    State
1   1       0       GE0/0/0   00e0fcdc5981  00e0fc9667e0  UP
```

可以观察到，会话状态正常（状态为 UP 表示正常）。

步骤三　测试各计算机的互通性

使用 PC1 计算机 ping ISP 路由器 LoopBack。

```
PC>ping 10.10.10.1

Ping 10.10.10.1: 32 data bytes, Press Ctrl_C to break
From 10.10.10.1: bytes=32 seq=1 ttl=254 time=47 ms
From 10.10.10.1: bytes=32 seq=2 ttl=254 time=47 ms
From 10.10.10.1: bytes=32 seq=3 ttl=254 time=31 ms
From 10.10.10.1: bytes=32 seq=4 ttl=254 time=47 ms
From 10.10.10.1: bytes=32 seq=5 ttl=254 time=47 ms

---10.10.10.1 ping statistics ---
  5 packet(s) transmitted
  5 packet(s) received
  0.00%packet loss
  round-trip min/avg/max = 31/43/47 ms
```

可以观察到，PC1 与 ISP 路由器之间通信正常。

项目小结

通过本项目的学习，同学们理解了广域网的基本概念，包括链路类型、相关协议等；理解了 PPP 的基本概念，通过项目操作掌握了 PPP 的两种身份认证的配置方法，即 PAP 和 CHAP；理解了 PPPoE 的基本概念，通过项目操作掌握了 PPPoE 的配置。

项目 8　网络安全技术

项目目标

1. 掌握 ACL 的配置方法
2. 掌握 NAT 的配置方法
3. 掌握本地 AAA 配置方法
4. 掌握 IPSec VPN 配置方法
5. 掌握 GRE 隧道配置方法

任务 1　ACL 技术

任务 1.1　使用基本 ACL 限制公司网络访问

任务背景

某公司有开发部、市场部和财务部,各有计算机若干台、财务系统服务器 1 台,使用三层交换机进行局域网组建,并通过路由器连接至外部网络。出于数据安全的考虑,该公司需要在交换机上进行访问控制。网络拓扑图如图 8-1 所示。具体要求如下:

(1) SW1 上为开发部、市场部、财务部及财务系统分别创建了 VLAN 10、VLAN 20、VLAN 30、VLAN 40;

(2) 要求财务系统服务器仅允许财务部访问;

(3) 财务系统服务器仅在内网使用,不允许访问外部网络;

(4) 测试计算机、交换机和路由器的 IP 地址及接口信息如图 8-1 所示。

使用基本 ACL 限制公司网络访问

任务规划

三层交换机的访问控制策略主要是通过 ACL 访问控制列表对不同 VLAN 的 IP 地址段进行流量匹配控制。标准 ACL 可以对 IP 包进行源地址匹配,即检查通过 IP 包中

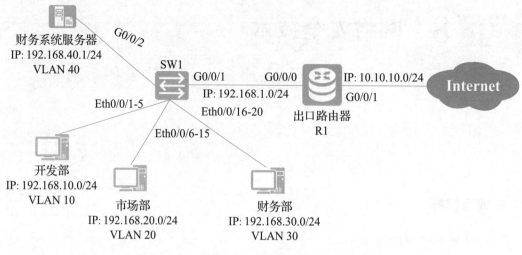

图 8-1 网络拓扑图

的源地址信息，如果源地址与 ACL 中的规则相匹配，就执行放行或拦截的操作。为了让其他部门无法访问财务系统服务器，我们可以在三层交换机中配置匹配财务部 IP 地址段、拒绝其他所有 IP 的 ACL，并在 G0/0/2 接口的 OUT 方向上应用；同时再添加拒绝财务部系统服务器 IP 地址段的 ACL，在 G0/0/1 接口的 OUT 方向上应用，阻止财务部系统服务器访问外部网络。外部网络连接方面，三层交换机配置默认路由指向出口路由器。出口路由器可根据 ISP 接入方式采用对应的路由协议，这里不作描述。

配置步骤如下：

（1）配置交换机基础环境；
（2）配置路由器基础环境；
（3）配置基本 ACL 访问控制；
（4）配置各部门计算机的 IP 地址。

具体规划见表 8-1、表 8-2、表 8-3。

表 8-1 VLAN 规划表

VLAN ID	IP 地址段	用途
VLAN 10	192.168.10.0/24	开发部
VLAN 20	192.168.20.0/24	市场部
VLAN 30	192.168.30.0/24	财务部
VLAN 40	192.168.40.0/24	财务系统
VLAN 50	192.168.1.0/24	连接外部网络

表 8-2 端口规划表

本端设备	端口号	端口类型	所属 VLAN	对端设备
SW1	Eth0/0/1-5	Access	VLAN 10	开发部 PC
SW1	Eth0/0/6-15	Access	VLAN 20	市场部 PC

续表

本端设备	端口号	端口类型	所属 VLAN	对端设备
SW1	Eth0/0/16-20	Access	VLAN 30	财务部 PC
SW1	G0/0/1	Access	VLAN 40	财务系统
SW1	G0/0/2	Access	VLAN 50	R1

表 8-3 IP 地址规划表

设备	接口	IP 地址
R1	G0/0/0	192.168.1.1/24
R1	G0/0/1	10.10.10.1/24
SW1	VLANIF 10	192.168.10.254/24
SW1	VLANIF 20	192.168.20.254/24
SW1	VLANIF 30	192.168.30.254/24
SW1	VLANIF 40	192.168.40.254/24
SW1	VLANIF 50	192.168.1.254/24
财务系统服务器		192.168.40.1/24
开发部		192.168.10.1/24
市场部		192.168.20.1/24
财务部		192.168.30.1/24

任务实施

步骤一 配置交换机基础环境

(1) 为各部门创建相应的 VLAN。

```
<Huawei>system-view
[Huawei]sysname SW1
[SW1]vlan batch 10 20 30 40 50
```

(2) 将各部门计算机所使用的端口类型转换为 Access 模式，并设置接口 PVID，将端口划分到相应的 VLAN。

```
[SW1]port-group group-member Ethernet 0/0/1 to Ethernet 0/0/5
[SW1-port-group]port link-type access
[SW1-port-group]port default vlan 10
[SW1-port-group]quit
[SW1]port-group group-member Ethernet 0/0/6 to Ethernet 0/0/15
[SW1-port-group]port link-type access
[SW1-port-group]port default vlan 20
[SW1-port-group] quit
```

```
[SW1]port-group group-member Ethernet 0/0/16 to Ethernet 0/0/20
[SW1-port-group]port link-type access
[SW1-port-group]port default vlan 30
[SW1-port-group] quit
[SW1]interface G0/0/2
[SW1-GigabitEthernet0/0/2]port link-type access
[SW1-GigabitEthernet0/0/2]port default vlan 40
[SW1-port-group] quit
[SW1]interface G0/0/1
[SW1-GigabitEthernet0/0/1]port link-type access
[SW1-GigabitEthernet0/0/1]port default vlan 50
[SW1-port-group] quit
```

（3）配置 VLANIF 接口的 IP 地址，作为各部门的网关。

```
[SW1]interface Vlanif 10
[SW1-Vlanif10]ip add 192.168.10.254 24
[SW1]interface Vlanif 20
[SW1-Vlanif20]ip add 192.168.20.254 24
[SW1]interface Vlanif 30
[SW1-Vlanif30]ip add 192.168.30.254 24
[SW1]interface Vlanif 40
[SW1-Vlanif40]ip add 192.168.40.254 24
[SW1]interface Vlanif 50
[SW1-Vlanif50]ip add 192.168.1.254 24
```

（4）配置交换机默认路由。

```
[SW1]ip route-static 0.0.0.0 0.0.0.0 192.168.1.1
```

步骤二 配置路由器基础环境

（1）配置路由器接口 IP 地址。

```
[Huawei]system-view
[Huawei]sysname R1
[R1]int G0/0/0
[R1-GigabitEthernet0/0/0]ip add 192.168.1.1 24
[R1]int G0/0/1
[R1-GigabitEthernet0/0/1]ip add 10.10.10.1 24
```

（2）配置路由器静态路由。

```
[R1]ip route-static 192.168.10.0 255.255.255.0 192.168.1.254
[R1]ip route-static 192.168.20.0 255.255.255.0 192.168.1.254
```

```
[R1]ip route-static 192.168.30.0 255.255.255.0 192.168.1.254
[R1]ip route-static 192.168.40.0 255.255.255.0 192.168.1.254
```

步骤三 配置基本 ACL 控制访问

(1) 在交换机上配置 ACL 规则，允许数据包源网段为 192.168.30.0 的报文通过。将规则应用到 G0/0/2 的端口上。

```
[SW1]acl 2000
[SW1-acl-basic-2000]rule permit source 192.168.30.0 0.0.0.255
[SW1-acl-basic-2000]rule deny
[SW1]int G0/0/2
[SW1-GigabitEthernet0/0/2]traffic-filter outbound acl 2000
```

(2) 在交换机上配置 ACL 规则，拒绝数据包源网段为 192.168.40.0 的报文通过。将规则应用到 G0/0/1 的端口上。

```
[SW1]acl 2001
[SW1-acl-basic-2001]rule deny source 192.168.40.0 0.0.0.255
[SW1]int G0/0/1
[SW1-GigabitEthernet0/0/1]traffic-filter outbound acl 2001
```

步骤四 配置各部门计算机的 IP 地址

各部门计算机的 IP 地址配置如图 8-2～图 8-5 所示。

图 8-2 财务系统服务器 IP 地址配置图

图 8-3 开发部 PC IP 地址配置图

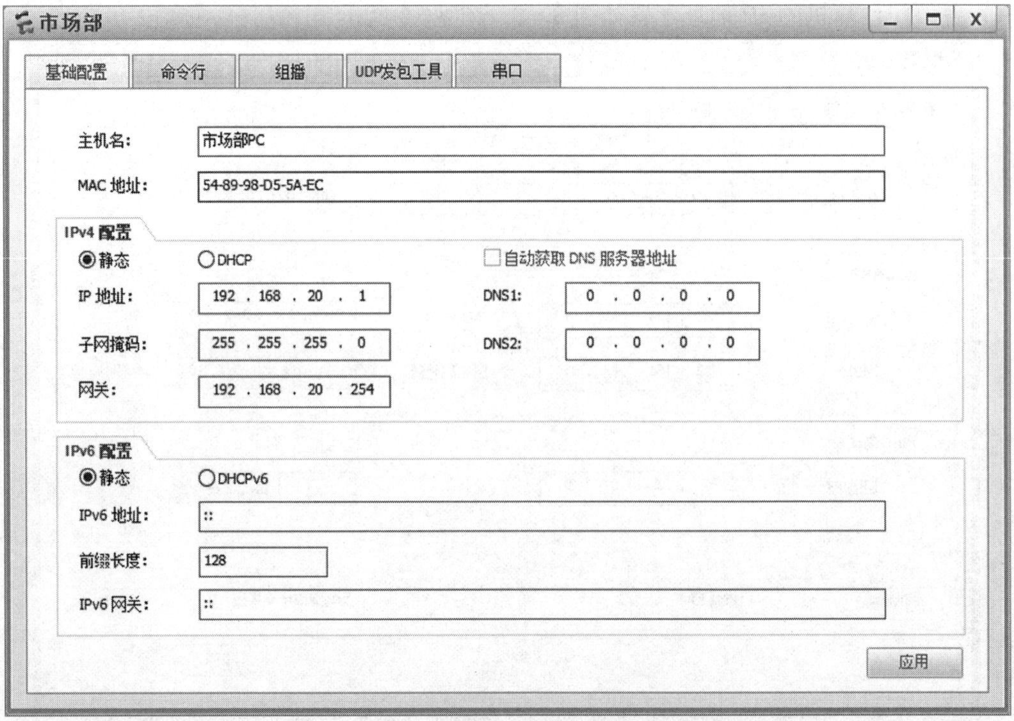

图 8-4 市场部 PC IP 地址配置图

图 8-5　财务部 PC IP 地址配置图

任务验证

步骤一　查看访问控制列表

SW1 的配置：

```
[SW1]display acl all
 Total nonempty ACL number is 2

Basic ACL 2000, 2 rules
Acl's step is 5
 rule 5 permit source 192.168.30.0 0.0.0.255
 rule 10 deny

Basic ACL 2001, 1 rule
Acl's step is 5
 rule 5 deny source 192.168.40.0 0.0.0.255
```

步骤二　测试各部门计算机的互通性

通过 ping 命令，测试各部门内部通信情况。

使用开发部的计算机 ping 市场部及财务部的计算机：

```
PC>ping 192.168.20.1

Ping 192.168.20.1: 32 data bytes, Press Ctrl_C to break
From 192.168.20.1: bytes=32 seq=1 ttl=127 time=47 ms
From 192.168.20.1: bytes=32 seq=2 ttl=127 time=47 ms
From 192.168.20.1: bytes=32 seq=3 ttl=127 time=31 ms
From 192.168.20.1: bytes=32 seq=4 ttl=127 time=31 ms
From 192.168.20.1: bytes=32 seq=5 ttl=127 time=31 ms
---192.168.20.1 ping statistics ---
  5 packet(s) transmitted
  5 packet(s) received
  0.00%packet loss
  round-trip min/avg/max = 31/37/47 ms

PC>ping 192.168.30.1

Ping 192.168.30.1: 32 data bytes, Press Ctrl_C to break
From 192.168.30.1: bytes=32 seq=1 ttl=127 time=32 ms
From 192.168.30.1: bytes=32 seq=2 ttl=127 time=31 ms
From 192.168.30.1: bytes=32 seq=3 ttl=127 time=47 ms
From 192.168.30.1: bytes=32 seq=4 ttl=127 time=31 ms
From 192.168.30.1: bytes=32 seq=5 ttl=127 time=31 ms
---192.168.30.1 ping statistics ---
  5 packet(s) transmitted
  5 packet(s) received
  0.00%packet loss
  round-trip min/avg/max = 31/34/47 ms
```

步骤三 测试各部门与财务系统的连接性

使用开发部的计算机 ping 财务系统服务器：

```
PC>ping 192.168.40.1

Ping 192.168.40.1: 32 data bytes, Press Ctrl_C to break
Request timeout!
Request timeout!
Request timeout!
Request timeout!
Request timeout!
---192.168.40.1 ping statistics ---
  5 packet(s) transmitted
  0 packet(s) received
  100.00%packet loss
```

使用财务部的计算机 ping 财务系统服务器：

```
PC>ping 192.168.40.1

Ping 192.168.40.1: 32 data bytes, Press Ctrl_C to break
From 192.168.40.1: bytes=32 seq=1 ttl=127 time=47 ms
From 192.168.40.1: bytes=32 seq=2 ttl=127 time=31 ms
From 192.168.40.1: bytes=32 seq=3 ttl=127 time=32 ms
From 192.168.40.1: bytes=32 seq=4 ttl=127 time=31 ms
From 192.168.40.1: bytes=32 seq=5 ttl=127 time=47 ms
---192.168.40.1 ping statistics ---
  5 packet(s) transmitted
  5 packet(s) received
  0.00%packet loss
  round-trip min/avg/max =31/37/47 ms
```

可以观察到，其他部门无法连接到财务系统服务器上，唯有财务部可以连接到财务系统服务器上。

步骤四 测试外部网络的连通性

通过 ping 命令，测试各部门 PC 及财务系统服务器能否访问外网。

使用开发部的计算机 ping 外部网络：

```
PC>ping 10.10.10.1

Ping 10.10.10.1: 32 data bytes, Press Ctrl_C to break
From 10.10.10.1: bytes=32 seq=1 ttl=254 time=31 ms
From 10.10.10.1: bytes=32 seq=2 ttl=254 time=47 ms
From 10.10.10.1: bytes=32 seq=3 ttl=254 time=31 ms
From 10.10.10.1: bytes=32 seq=4 ttl=254 time=47 ms
From 10.10.10.1: bytes=32 seq=5 ttl=254 time=31 ms
---10.10.10.1 ping statistics ---
  5 packet(s) transmitted
  5 packet(s) received
  0.00%packet loss
  round-trip min/avg/max =31/37/47 ms
```

使用财务系统服务器 ping 外部网络：

```
PC>ping 10.10.10.1

Ping 10.10.10.1: 32 data bytes, Press Ctrl_C to break
Request timeout!
Request timeout!
Request timeout!
```

```
Request timeout!
Request timeout!
---10.10.10.1 ping statistics ---
  5 packet(s) transmitted
  0 packet(s) received
  100.00%packet loss
```

可以观察到，其他部门均能访问外部网络，唯有财务系统服务器无法访问外部网络。

任务 1.2　使用扩展 ACL 限制公司网络访问

任务背景

某公司财务部有计算机若干台，并架设了专用的财务系统服务器，进行局域网组建，通过路由器连接至互联网。出于财务系统数据安全的考虑，该公司需要在路由器上配置访问控制策略，只有财务部 PC1 能够访问财务系统前端网站；同时，服务器不可访问外部网络。网络拓扑图如图 8-6 所示。具体要求如下：

（1）仅允许财务部 PC1 访问财务系统服务器前端网站；
（2）财务系统服务器仅在内网使用，不允许访问外部网络；
（3）测试计算机、路由器的 IP 地址及接口信息如图 8-6 所示。

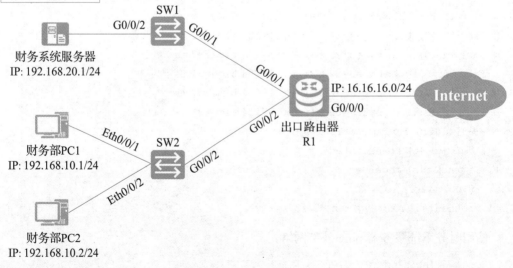

图 8-6　网络拓扑图

任务规划

扩展 ACL 可以对 IP 包地址信息中的源地址、目的地址、协议、端口号进行匹配，即检查通过 IP 包中的地址信息，如果地址信息与 ACL 中的规则相匹配，就执行放行或拦截的操作。在本项目中，访问控制策略主要集中在对财务系统服务器的访问权限上，通过路由器上的应用即可实现。ACL 策略的主要内容包括：（1）创建允许财务部 PC1 对服务器 80 端口的访问；（2）创建拒绝财务系统服务器对外部网络的访问。

配置步骤如下：

（1）配置路由器接口；

（2）配置高级 ACL 控制访问；

（3）配置各部门计算机的 IP 地址。

具体规划见表 8-4。

表 8-4　IP 地址规划表

设备	接口	IP 地址
R1	G0/0/0	16.16.16.16/24
R1	G0/0/1	192.168.20.254/24
R1	G0/0/2	192.168.10.254/24
财务系统服务器		192.168.20.1/24
财务部 PC1		192.168.10.1/24
财务部 PC2		192.168.10.2/24

任务实施

步骤一　配置路由器接口

R1 的配置：

```
<Huawei>system-view
[Huawei]sysname R1
[R1]interface GigabitEthernet 0/0/0
[R1-GigabitEthernet0/0/0]ip address 16.16.16.16 255.255.255.0
[R1]interface GigabitEthernet 0/0/1
[R1-GigabitEthernet0/0/1]ip address 192.168.20.254 255.255.255.0
[R1]interface GigabitEthernet 0/0/2
[R1-GigabitEthernet0/0/2]ip address 192.168.10.254 255.255.255.0
```

步骤二　配置高级 ACL 控制访问

（1）在路由器上配置高级 ACL 规则，允许财务部 PC1 对服务器 80 端口的访问。将规则应用到 G0/0/1 的端口上。

```
[R1]acl 3000
[R1-acl-adv-3000]rule 5 permit tcp source 192.168.10.1 0 destination-port eq www
[R1-acl-adv-3000]rule 10 deny ip
[R1]int G0/0/1
[R1-GigabitEthernet0/0/1]traffic-filter outbound acl 3000
```

(2) 在路由器上配置高级 ACL 规则，拒绝财务系统服务器对外网的访问。将规则应用到 G0/0/0 的端口上。

```
[R1]acl 3001
[R1-acl-adv-3001]rule 5 deny ip source 192.168.20.0 0.0.0.255
[R1-acl-adv-3001]rule 10 permit ip
[R1]int g0/0/0
[R1-GigabitEthernet0/0/1]traffic-filter outbound acl 3001
```

步骤三 配置各部门计算机的 IP 地址

各部门计算机的 IP 地址配置如图 8-7～图 8-9 所示。

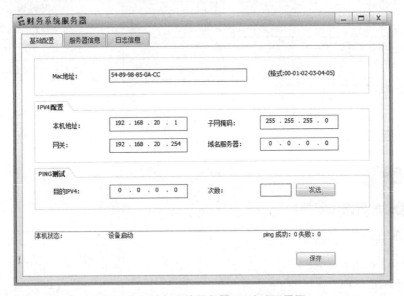

图 8-7 财务系统服务器 IP 地址配置图

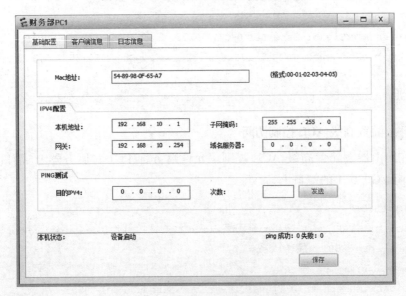

图 8-8 财务部 PC1 IP 地址配置图

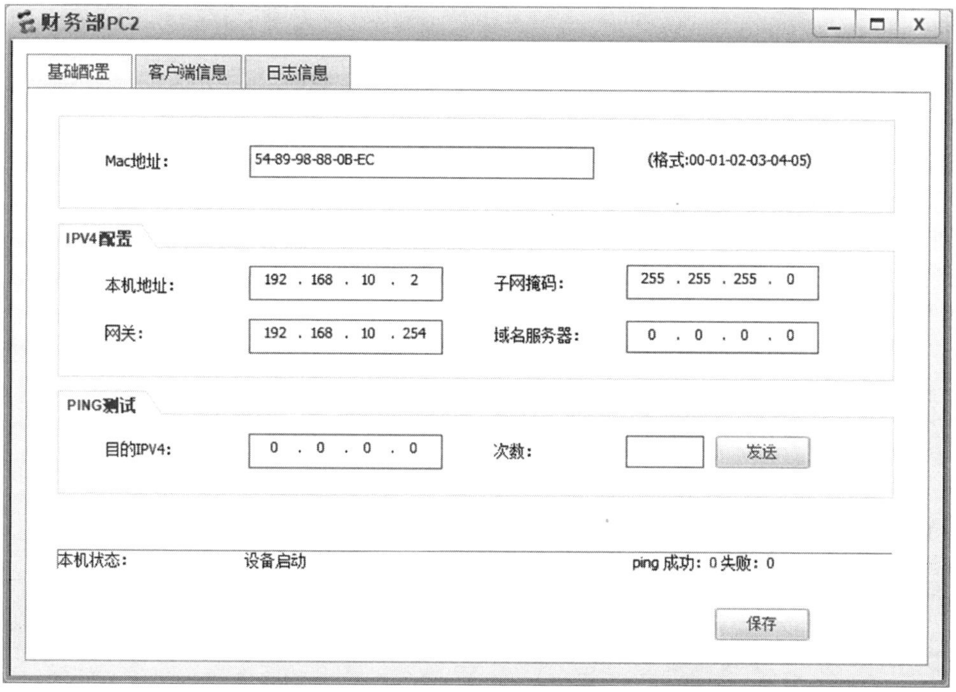

图 8-9 财务部 PC2 IP 地址配置图

任务验证

步骤一 查看访问控制列表

R1 的配置：

```
[R1]dis acl all
 Total quantity of nonempty ACL number is 2

Advanced ACL 3000, 2 rules
Acl's step is 5
 rule 5 permit tcp source 192.168.10.1 0 destination-port eq www
 rule 10 deny ip (2 matches)

Advanced ACL 3001, 1 rule
Acl's step is 5
 rule 5 deny tcp source 192.168.20.1 0
```

步骤二 测试财务部 PC 与服务器 80 端口的连通性

（1）配置财务系统服务器 HttpServer，启用 80 端口，如图 8-10 所示。

图 8-10 财务系统服务器 HttpServer 配置图

（2）使用财务部 PC1 访问 Http，如图 8-11 所示。

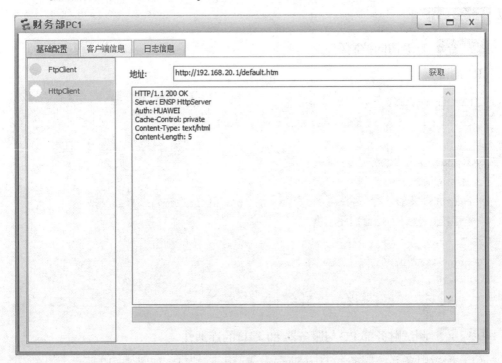

图 8-11 财务部 PC1 访问 Http

（3）使用财务部 PC2 访问 Http，如图 8-12 所示。

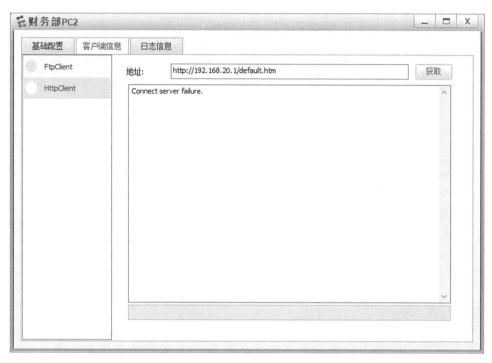

图 8-12　财务部 PC2 访问 Http

步骤三　测试服务器与外部网络的连通性

（1）使用财务系统服务器 ping 测试外部网络，如图 8-13 所示。

图 8-13　财务系统服务器 ping 测试外部网络

可以看出，财务部系统服务器不可以与外部网络通信。

（2）使用财务部 PC1 ping 测试外部网络，如图 8-14 所示。

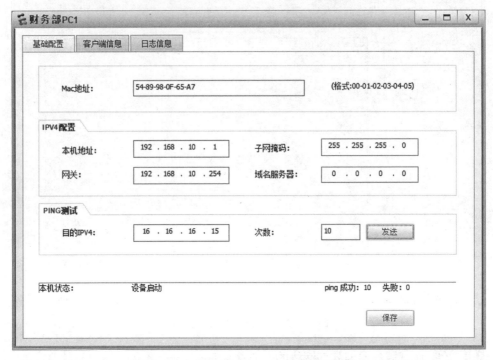

图 8-14　财务部 PC1 ping 测试外部网络

可以看出，财务部可以与外部网络通信。

任务 2　NAT 技术

任务 2.1　使用静态 NAT 对外发布公司官网

使用静态 NAT 对外发布公司官网

📖 **任务背景**

某公司搭建了网站服务器，用于对外发布公司官网。为了保障内部网络的安全和解决私有地址在公网上通信的问题，该公司需要在出口路由中配置 NAT，使内部服务器映射到公网地址上。

网络拓扑图如图 8-15 所示，具体要求如下：

（1）公司内网使用 192.168.1.0/24 网段，出口使用 16.16.16.0/24 网段；

（2）出口路由器上申请了一个 16.16.16.1 的 IP 可供网站服务器做 NAT 映射；

（3）测试计算机和路由器的 IP 地址及接口信息如图 8-15 所示。

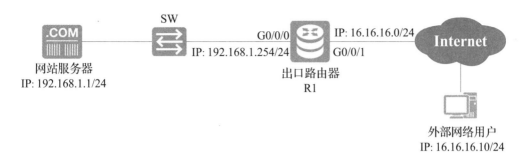

图 8-15 网络拓扑图

任务规划

通过 NAT 映射内部服务器需要使用专用的公网 IP 地址,故需要申请 2 个以上公网 IP 地址,一个用于服务器映射,一个用于内网的通信,这里使用 16.16.16.1 和 16.16.16.16 作为公网 IP 地址。网站服务器处于内部网络,IP 地址为 192.168.1.1,需要在出口路由器配置静态 NAT,将公网地址 16.16.16.1 实现一对一映射,这样即可通过公网地址直接访问内网的服务器。互联网连接方面,出口路由器可根据 ISP 服务商的网络环境配置相应的路由协议。

配置步骤如下:

(1)配置路由器接口;

(2)配置静态 NAT;

(3)配置各部门计算机的 IP 地址。

具体规划见表 8-5。

表 8-5 IP 地址规划表

设备	接口	IP 地址
R1	G0/0/0	192.168.1.254/24
R1	G0/0/1	16.16.16.16/24
网站服务器		192.168.1.1/24
外部网络用户		16.16.16.10/24

任务实施

步骤一 配置路由器接口

在路由器接口配置对应 IP。

```
[Huawei]system-view
[Huawei]sysname R1
[R1]interface GigabitEthernet 0/0/0
[R1-GigabitEthernet0/0/0]ip add 192.168.1.254 24
[R1]interface GigabitEthernet 0/0/1
[R1-GigabitEthernet0/0/1]ip add 16.16.16.16 24
```

步骤二 配置静态 NAT

在 R1 的 G0/0/1 接口下使用 nat static 命令配置内部地址到外部地址一对一转换。

```
[R1]interface GigabitEthernet 0/0/1
[R1-GigabitEthernet0/0/1]nat static global 16.16.16.1 inside 192.168.1.1
```

步骤三 配置各部门计算机的 IP 地址

各部门计算机的 IP 地址配置如图 8-16、图 8-17 所示。

图 8-16 服务器 IP 地址配置图

图 8-17 外部网络用户 IP 地址配置图

项目 8 网络安全技术

任务验证

步骤一 验证 NAT 静态配置信息

R1 的配置：

```
<R1>display nat static
  Static Nat Information:
  Interface  : GigabitEthernet0/0/1
    Global IP/Port     : 16.16.16.1/----
    Inside IP/Port     : 192.168.1.1/----
    Protocol : ----
    VPN instance-name  : ----
    Acl number         : ----
    Netmask : 255.255.255.255
    Description : ----

  Total : 1
```

步骤二 互联网 PC 测试与网站服务器的连通性

（1）外部网络用户使用 ping 命令测试网站服务器的互联网 IP 能否访问：

```
PC>ping 16.16.16.1

Ping 16.16.16.1: 32 data bytes, Press Ctrl_C to break
From 16.16.16.1: bytes=32 seq=1 ttl=127 time=47 ms
From 16.16.16.1: bytes=32 seq=2 ttl=127 time=47 ms
From 16.16.16.1: bytes=32 seq=3 ttl=127 time=47 ms
From 16.16.16.1: bytes=32 seq=4 ttl=127 time=46 ms
From 16.16.16.1: bytes=32 seq=5 ttl=127 time=47 ms

---16.16.16.1 ping statistics ---
  5 packet(s) transmitted
  5 packet(s) received
  0.00% packet loss
  round-trip min/avg/max = 46/46/47 ms
```

（2）在 R1 上使用 dis nat session all 命令查看 NAT 转换信息：

```
[R1]dis nat session all
  NAT Session Table Information:

    Protocol          : ICMP(1)
    SrcAddr   Vpn     : 16.16.16.10
    DestAddr  Vpn     : 16.16.16.1
```

```
        Type Code IcmpId  : 0   8    31521
        NAT-Info
          New SrcAddr     : ----
          New DestAddr    : 192.168.1.1
          New IcmpId      : ----

        Protocol          : ICMP(1)
        SrcAddr    Vpn    : 16.16.16.10
        DestAddr   Vpn    : 16.16.16.1
        Type Code IcmpId  : 0   8    31520
        NAT-Info
          New SrcAddr     : ----
          New DestAddr    : 192.168.1.1
          New IcmpId      : ----
```

可以看到，R1 收到互联网访问目的地址为 16.16.16.1 的流量时，将目的地址转换为 192.168.1.1，使其能正常访问服务器。

任务 2.2　某公司使用动态 NAT 访问互联网

📖 任务背景

某公司有计算机若干台，利用交换机组建了局域网，并通过出口路由器连接互联网。为了保障内部网络的安全和解决私有地址在互联网上通信的问题，该公司需要在出口路由器中配置动态 NAT，使内部计算机 IP 地址映射为公网 IP 地址访问互联网。网络拓扑图如图 8-18 所示，具体要求如下：

（1）公司内网使用 192.168.10.0/24 网段，出口为 16.16.16.0/24 网段；

（2）出口路由器上申请了 16.16.16.1-5 等互联网 IP 地址，可供 NAT 转换使用；

（3）测试计算机和路由器的 IP 地址及接口信息如图 8-18 所示。

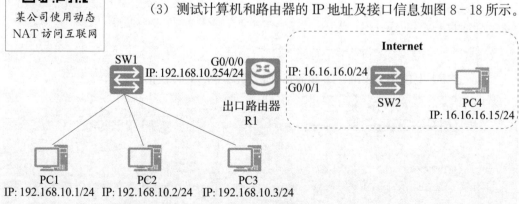

图 8-18　网络拓扑图

任务规划

动态 NAT 转换需要有多个公网 IP 地址，这里以 16.16.16.1-5 作为转换后的公网 IP 地址。在路由器中将公网 IP 地址配置为 NAT 地址池，并建立 ACL 列表匹配内部地址，在出口路由器的 G0/0/1 上应用 NAT 转换即可。互联网连接方面，出口路由器可根据 ISP 服务商的网络环境配置相应的路由协议。

配置步骤如下：

（1）配置路由器接口；

（2）配置动态 NAT；

（3）配置各计算机的 IP 地址。

具体规划见表 8-6。

表 8-6　IP 地址规划表

设备	接口	IP 地址
R1	G0/0/0	192.168.10.254/24
R1	G0/0/1	16.16.16.16/24
PC1	E0/0/1	192.168.10.1/24
PC2	E0/0/1	192.168.10.2/24
PC3	E0/0/1	192.168.10.3/24
PC4	E0/0/1	16.16.16.15/24

任务实施

步骤一　配置路由器接口

在路由器接口配置对应的 IP。

```
[Huawei]system-view
[Huawei]sysname R1
[R1]interface GigabitEthernet 0/0/0
[R1-GigabitEthernet0/0/0]ip add 192.168.10.254 24
[R1]interface GigabitEthernet 0/0/1
[R1-GigabitEthernet0/0/1]ip add 16.16.16.16 24
```

步骤二　配置动态 NAT

在 R1 上使用 nat address-group 命令配置 NAT 地址池，设置起始和结束地址分别为 16.16.16.1 和 16.16.16.5。

```
[R1]nat address-group 1 16.16.16.1 16.16.16.5
```

创建基本 ACL2000。

```
[R1]acl 2000
[R1-acl-basic-2000]rule permit source 192.168.10.0 0.0.0.255
```

在 G0/0/1 接口下使用 nat outbound 命令将 ACL2000 与地址池相关联，使得 ACL 中规定的地址可以使用地址池进行地址转换。

```
[R1-acl-basic-2000]int G0/0/1
[R1-GigabitEthernet0/0/1]nat outbound 2000 address-group 1 no-pat
```

步骤三 配置各计算机的 IP 地址

各计算机的 IP 地址配置如图 8-19～图 8-22 所示。

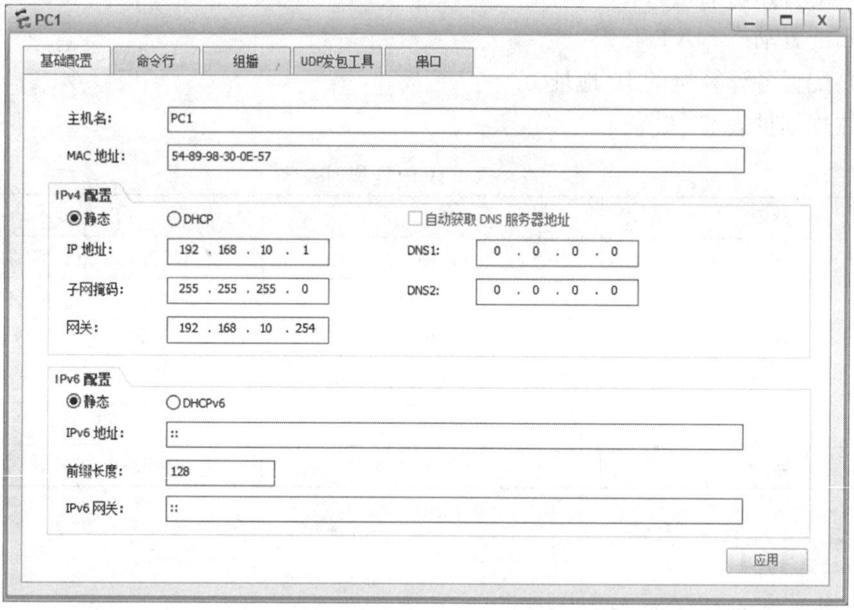

图 8-19　PC1 IP 地址配置图

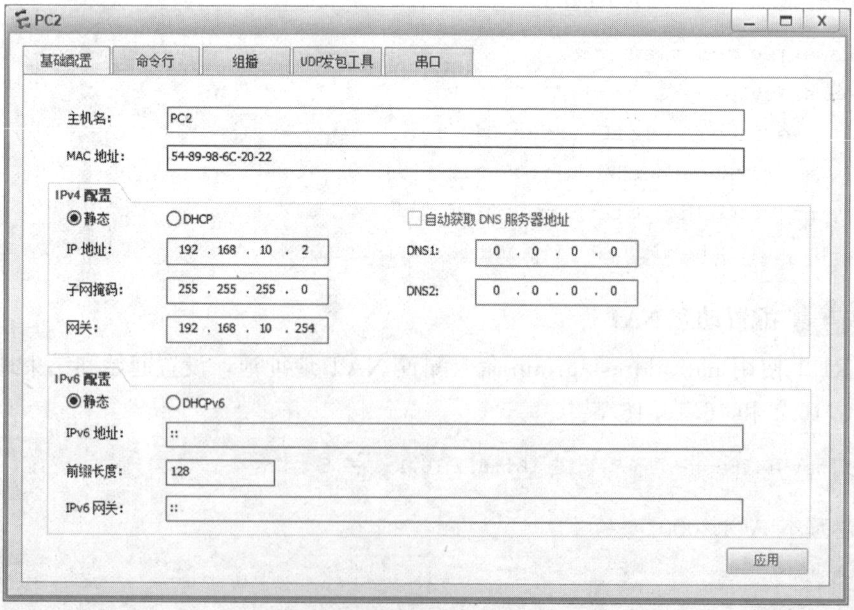

图 8-20　PC2 IP 地址配置图

图 8-21　PC3 IP 地址配置图

图 8-22　PC4 IP 地址配置图

任务验证

步骤一　**查看 NAT Outbound 信息**

在 R1 上使用 display nat outbound 命令查看 NAT Outbound 信息。

```
<R1>display nat outbound
NAT Outbound Information:
--------------------------------------------------------------------------
 Interface                Acl        Address-group/IP/Interface    Type
--------------------------------------------------------------------------
 GigabitEthernet0/0/1     2000                1                    no-pat
--------------------------------------------------------------------------
 Total : 1
```

步骤二 测试与互联网的连通性

使用 PC1 测试与互联网的连通性，并在 R1 的接口 G0/0/1 上抓包。

```
PC>ping 16.16.16.15

Ping 16.16.16.15: 32 data bytes, Press Ctrl_C to break
From 16.16.16.15: bytes=32 seq=1 ttl=127 time=31 ms
From 16.16.16.15: bytes=32 seq=2 ttl=127 time=31 ms
From 16.16.16.15: bytes=32 seq=3 ttl=127 time=31 ms
From 16.16.16.15: bytes=32 seq=4 ttl=127 time=32 ms
From 16.16.16.15: bytes=32 seq=5 ttl=127 time=31 ms

---16.16.16.15 ping statistics ---
  5 packet(s) transmitted
  5 packet(s) received
  0.00%packet loss
  round-trip min/avg/max =31/31/32 ms
```

步骤三 查看 NAT 会话信息

在 R1 上使用 display nat session 命令查看 NAT 会话信息。

```
<R1>display nat session all
  NAT Session Table Information:

    Protocol          : ICMP(1)
    SrcAddr   Vpn     : 192.168.10.1
    DestAddr  Vpn     : 16.16.16.15
    Type Code IcmpId  : 0   8   33944
    NAT-Info
      New SrcAddr     : 16.16.16.1
      New DestAddr    : ----
      New IcmpId      : ----

    Protocol          : ICMP(1)
    SrcAddr   Vpn     : 192.168.10.1
    DestAddr  Vpn     : 16.16.16.15
```

```
            Type Code IcmpId  : 0   8   33946
        NAT-Info
          New SrcAddr        : 16.16.16.2
          New DestAddr       : ----
          New IcmpId         : ----

          Protocol           : ICMP(1)
          SrcAddr    Vpn     : 192.168.10.1
          DestAddr   Vpn     : 16.16.16.15
          Type Code IcmpId   : 0   8   33947
```

可以看到，R1 收到源为 192.168.10.1 的流量时，将其 NAT 转换源地址为互联网地址 16.16.16.1，使其能正常访问互联网。

任务 2.3 使用静态 NAPT 对外发布公司官网

📋 任务背景

某公司搭建了网站服务器，用于对外发布官方网站，公司只租用了一个互联网 IP 地址用于互联网的访问。为了保障内部网络的安全和解决互联网 IP 地址不足的问题，该公司需要在出口路由中配置静态 NAPT，用于将内部服务器映射到互联网 IP 地址上。网络拓扑图如图 8-23 所示，具体要求如下：

使用静态 NAPT 对外发布公司官网

（1）公司内网使用 192.168.1.0/24 网段，出口使用 16.16.16.0/24 网段；

（2）因为公司仅申请了一个互联网 IP，需要配置静态 NAPT，所以仅将网站服务器的 80 端口做映射；

（3）测试计算机和路由器的 IP 地址及接口信息如图 8-23 所示。

图 8-23　网络拓扑图

🛠 任务规划

在只有一个互联网 IP 地址的情况下进行内部服务对外映射，需采用静态 NAPT 的方式。静态 NAPT 是通过 IP 地址和端口对应映射的方式，将内部服务器的某一服务发布到互联网上的。出口路由器的 G0/0/1 的 IP 地址为 16.16.16.1/24，通过配置静态 NAPT，将内部服务器的 80 端口对应映射到 G0/0/1 接口 IP 地址上的 80 端口，即可实现对外发布服务。互联网连接方面，出口路由器可根据 ISP 服务商的网络环境配置相应的路由协议。

配置步骤如下：
（1）配置路由器接口；
（2）配置静态 NAPT；
（3）配置各计算机的 IP 地址。

具体规划见表 8-7。

表 8-7 IP 地址规划表

设备	接口	IP 地址
R1	G0/0/0	192.168.1.254/24
R1	G0/0/1	16.16.16.1/24
网站服务器		192.168.1.1/24
外部网络用户		16.16.16.10/24

任务实施

步骤一 配置路由器接口

在路由器接口配置对应的 IP。

```
[Huawei]sysname R1
[R1]int G0/0/0
[R1-GigabitEthernet0/0/0]ip add 192.168.1.254 24
[R1]int G0/0/1
[R1-GigabitEthernet0/0/1]ip add 16.16.16.1 24
```

步骤二 配置静态 NAPT

在 R1 的 G0/0/1 接口上，使用 nat server 命令定义内部服务器的映射表，指定服务器通信协议为 TCP，配置服务器使用的互联网 IP 为 16.16.16.1，服务器内网为 192.168.1.1，指定端口号为 80。

```
[R1]interface GigabitEthernet 0/0/1
[R1-GigabitEthernet0/0/1]nat server protocol tcp global 16.16.16.1 80 inside 192.168.1.1 80
```

步骤三 配置各计算机的 IP 地址

各计算机的 IP 地址配置如图 8-24、图 8-25 所示。

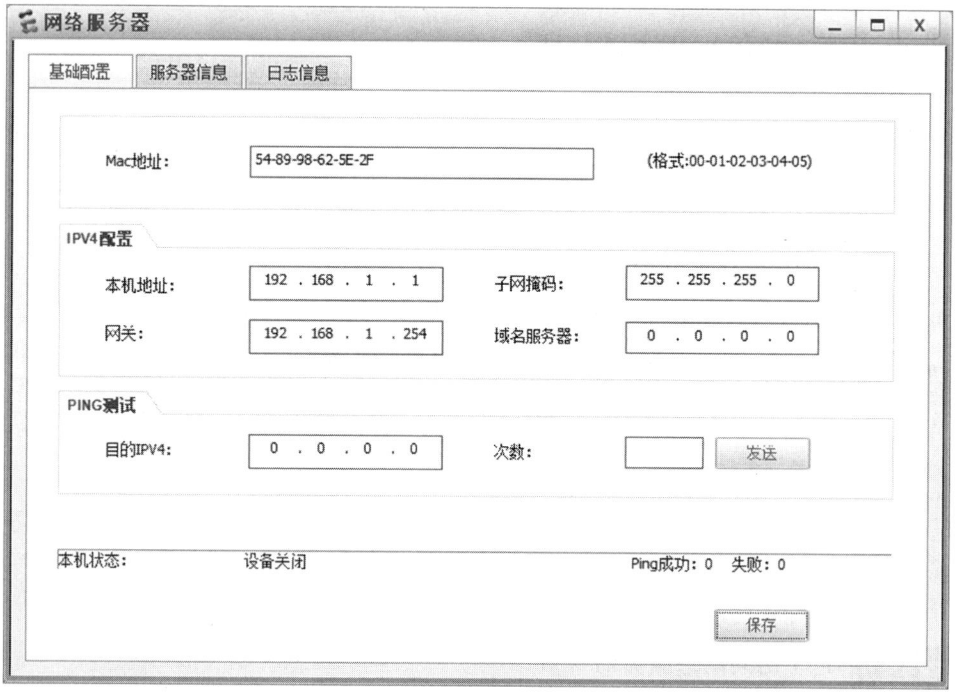

图 8－24　网络服务器 IP 地址配置

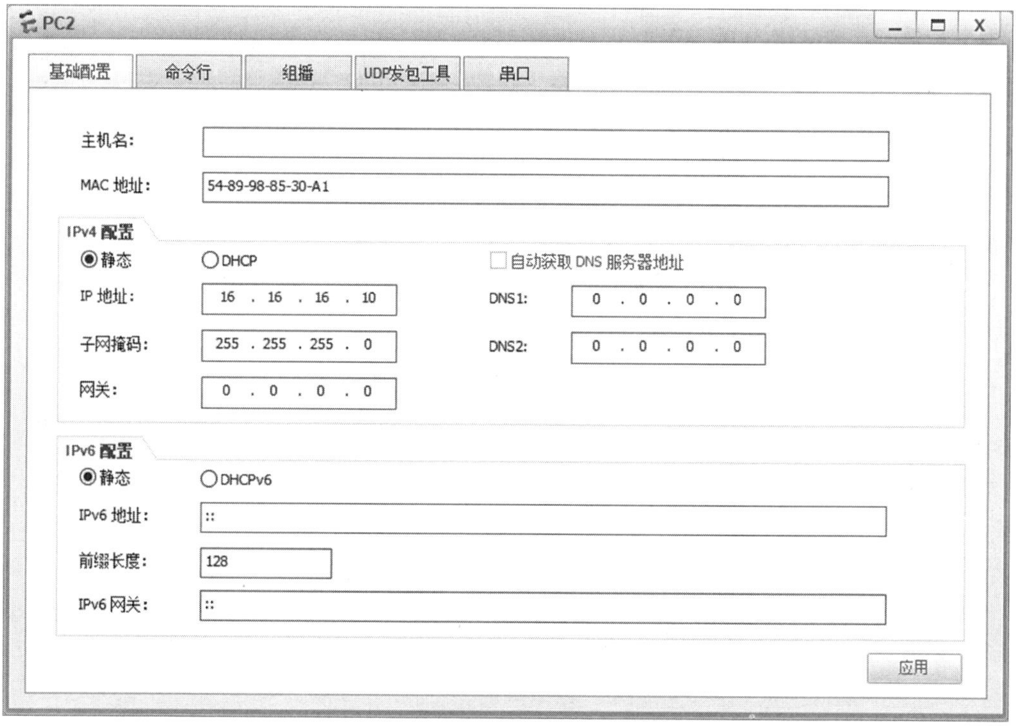

图 8－25　外部网络用户 IP 地址配置

任务验证

步骤一 验证静态 NAPT 配置信息

R1 的配置：

```
[R1]display nat server

  Nat Server Information:
  Interface  : GigabitEthernet0/0/1
    Global IP/Port    : 16.16.16.1/80(www)
    Inside IP/Port    : 192.168.1.1/80(www)
    Protocol : 6(tcp)
    VPN instance-name : ----
    Acl number        : ----
    Description : ----
Total :    1
```

可以观察到配置已经生效。

步骤二 PC1 访问网站服务器 web

网站服务器配置 HttpServer 如图 8-26 所示。

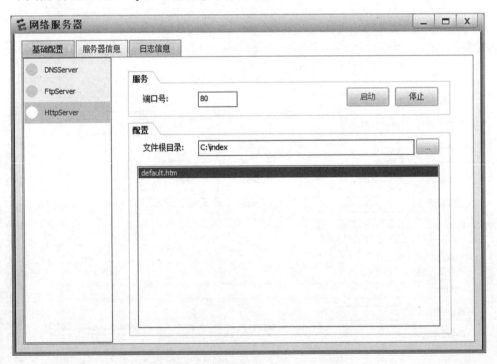

图 8-26 网络服务器 HttpServer 配置

在 PC1 访问网站服务器 web，如图 8-27 所示。

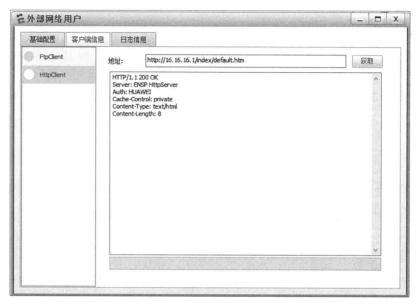

图 8-27 访问 web

可以观察到，PC 用户可以成功访问网站服务器。

任务 2.4　某公司使用 Easy IP 方式访问互联网

任务背景

某公司有计算机若干台，利用交换机组建了局域网，并通过出口路由器连接互联网。因业务开展的需要，该公司申请了一个公网 IP 地址，现需配置出口路由器，实现内部网络访问互联网。网络拓扑图如图 8-28 所示，具体要求如下：

（1）公司内网使用 192.168.10.0/24 网段，出口使用 16.16.16.0/24 网段；

（2）出口路由器上配置 Easy IP，使内部计算机可以访问互联网；

（3）测试计算机和路由器的 IP 地址及接口信息如图 8-28 所示。

微课视频

某公司使用 Easy IP 方式访问互联网

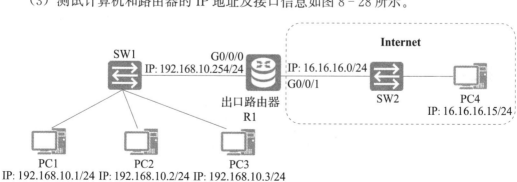

图 8-28　网络拓扑图

🖥️ 任务规划

Easy IP 是 NAT 的方式之一，主要用于内部计算机共享公网 IP 地址访问互联网。本项目中，出口路由器 G0/0/1 接口的 IP 地址为 16.16.16.16/24，通过创建 ACL 列表，匹配内部计算机的 IP 地址段，在出口路由器的 G0/0/1 接口上进行 Easy IP 的 NAT 转换，即可实现共享上网。互联网连接方面，出口路由器可根据 ISP 服务商的网络环境配置相应的路由协议。

配置步骤如下：
(1) 配置路由器接口；
(2) 配置动态 NAT；
(3) 配置各计算机的 IP 地址。

具体规划见表 8-8。

表 8-8　IP 地址规划表

设备	接口	IP 地址
R1	G0/0/0	192.168.10.254/24
R1	G0/0/1	16.16.16.16/24
PC1	E0/0/1	192.168.10.1/24
PC2	E0/0/1	192.168.10.2/24
PC3	E0/0/1	192.168.10.3/24
PC4	E0/0/1	16.16.16.15/24

🛠️ 任务实施

步骤一　配置路由器接口

在路由器接口配置对应的 IP。

```
[Huawei]sysname R1
[R1]int G0/0/0
[R1-GigabitEthernet0/0/0]ip add 192.168.10.254 24
[R1]int G0/0/1
[R1-GigabitEthernet0/0/1]ip add 16.16.16.16 24
```

步骤二　配置动态 NAT

创建基本 ACL2000。

```
[R1]acl 2000
[R1-acl-basic-2000]rule permit source 192.168.10.0 0.0.0.255
```

在 G0/0/1 接口下使用 nat outbound 命令配置 Easy-IP 特性，直接使用接口 IP 地址作为 NAT 转换后的地址。

```
[R1-acl-basic-2000]int G0/0/1
[R1-GigabitEthernet0/0/1]nat outbound 2000
```

步骤三 配置各计算机的 IP 地址

各计算机的 IP 地址配置如图 8-29～图 8-32 所示。

图 8-29　PC1 IP 地址配置图

图 8-30　PC2 IP 地址配置图

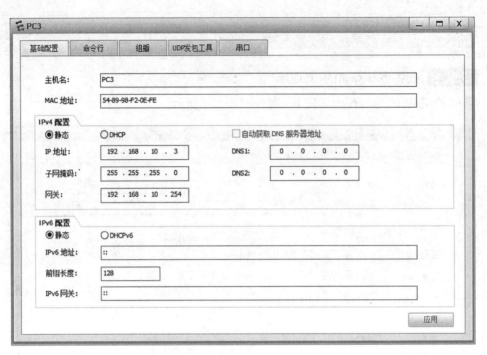

图 8-31 PC3 IP 地址配置图

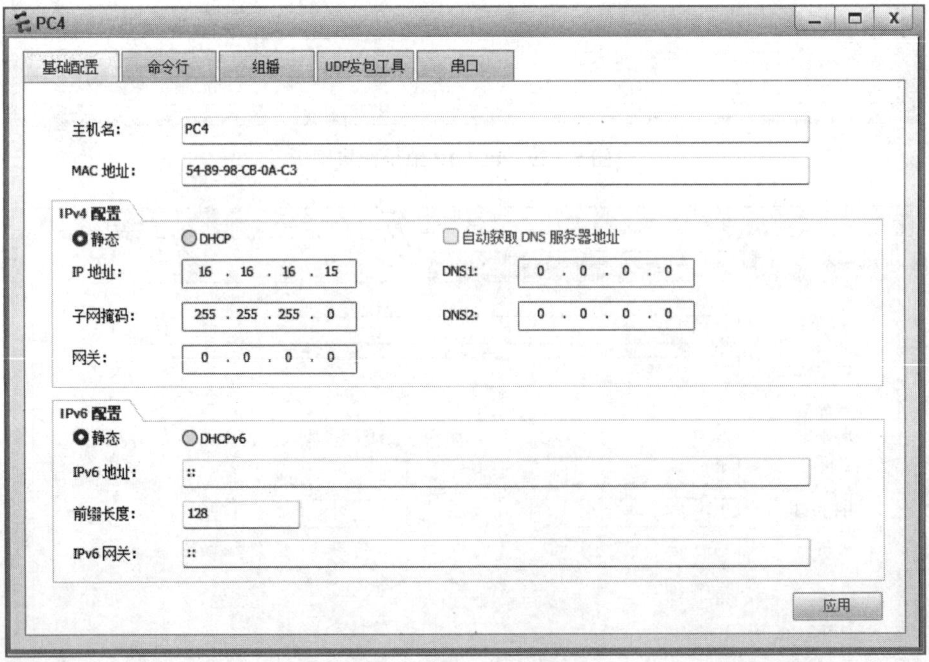

图 8-32 PC4 IP 地址配置图

任务验证

验证 NAT Session 信息：

在 PC1 和 PC2 上使用 UDP 发包工具发送 UDP 数据包到公网地址 16.16.16.10，

配置好目的 IP 和 UDP 源、目的端口号后，输入字符串数据后单击"发送"按钮，如图 8-33、图 8-34 所示。

图 8-33　PC1-UDP 发包配置

图 8-34　PC2-UDP 发包配置

在 PC1、PC2 发送 UDP 数据包后，在 R1 上查看 NAT Session 的详细信息。

```
[R1]display nat session protocol udp verbose
  NAT Session Table Information:

    Protocol            : UDP(17)
    SrcAddr   Port Vpn  : 192.168.10.2     2560
    DestAddr  Port Vpn  : 16.16.16.15      2560
    Time To Live        : 120 s
    NAT-Info
      New SrcAddr       : 16.16.16.16
      New SrcPort       : 10242
      New DestAddr      : ----
      New DestPort      : ----

    Protocol            : UDP(17)
    SrcAddr   Port Vpn  : 192.168.10.1     2560
    DestAddr  Port Vpn  : 16.16.16.15      2560
    Time To Live        : 120 s
    NAT-Info
      New SrcAddr       : 16.16.16.16
      New SrcPort       : 10243
      New DestAddr      : ----
      New DestPort      : ----

 Total : 2
```

可以观察到，源地址为 192.168.10.2 的 UDP 数据包被新源地址 16.16.16.16 和新端口 10242 替换，源地址为 192.168.10.1 的 UDP 数据包被新源地址 16.16.16.16 和新端口 10243 替换。R1 借用自身 G0/0/1 接口的公网 IP 地址为所有私网地址做 NAT 转换，使用不同的端口号区分不同的私网数据。

任务 3　访问控制技术

任务 3.1　本地 AAA 配置

任务背景

AAA 是一种提供 Authentication（认证）、Authorization（授权）和 Accounting（计费）的安全技术。该技术可以用于验证用户账户是否合法，授权用户可以访问的服务,并记录用户使用网络资源的情况。

用户可以使用 AAA 提供的一种或多种安全服务。例如,公司仅仅想让员工在访问某些特定资源的时候进行身份认证,那么网络管理员只要配置认证服务器即可;但是若希望对员工使用网络的情况进行记录,那么还需要配置计费服务器。

如上所述,AAA 是一种管理框架,它提供了授权部分用户去访问特定资源,同时可以记录这些用户操作行为的一种安全机制。AAA 因为具有良好的可扩展性,并且容易实现用户信息的集中管理,所以被广泛使用。AAA 可以通过多种协议来实现,在实际应用中,最常使用 RADIUS 协议。

本地 AAA 配置

任务规划

您是企业的网络管理员,需要对企业服务器的资源访问进行控制。由于只有通过认证的用户才能访问特定的资源,因此您需要在 R1 和 R3 两台路由器上配置本地 AAA 认证,并基于域来对用户进行管理,配置已认证用户的权限级别。本地 AAA 配置实验拓扑图如图 8-35 所示。

图 8-35 本地 AAA 配置实验拓扑图

任务实施

步骤一　实验环境准备

在路由器接口配置对应的 IP。

```
[Huawei]sysname R1
[R1]interface GigabitEthernet0/0/0
[R1-GigabitEthernet0/0/0]ip address 119.84.111.1 24

[Huawei]sysname R3
[R3]inter GigabitEthernet0/0/0
[R3-GigabitEthernet0/0/0]ip address 119.84.111.3 24
```

步骤二　在 R1 上配置 AAA 功能

(1) 在 R1 上配置认证方案为本地认证,授权方案为本地授权。

```
[R1]aaa
[R1-aaa]authentication-scheme auth1
Info: Create a new authentication scheme.
```

```
[R1-aaa-authen-auth1]authentication-mode local
[R1-aaa-authen-auth1]quit
[R1-aaa]authorization-scheme auth2
Info: Create a new authorization scheme.
[R1-aaa-author-auth2]authorization-mode local
[R1-aaa-author-auth2]quit
```

（2）在 R1 上创建域"huawei"并将认证方案和授权方案与域关联起来，然后创建一个用户并将用户加入域"huawei"。

```
[R1]telnet server enable
[R1]aaa
[R1-aaa]domain huawei
[R1-aaa-domain-huawei]authentication-scheme auth1
[R1-aaa-domain-huawei]authorization-scheme auth2
[R1-aaa-domain-huawei]quit
[R1-aaa]local-user user1@huawei password cipher huawei123
[R1-aaa]local-user user1@huawei service-type telnet
[R1-aaa]local-user user1@huawei privilege level 0
```

（3）将 R1 配置为 Telnet 服务器，并将认证模式配置为 AAA。

```
[R1]user-interface vty 0 4
[R1-ui-vty0-4]authentication-mode aaa
```

（4）验证 R3 Telnet R1 时是否要经过 AAA 认证。

```
<R3>telnet 119.84.111.1
  Press CTRL_] to quit telnet mode
  Trying 119.84.111.1 ...
  Connected to 119.84.111.1 ...

Login authentication

Username:user1@huawei
Password:
<R1>system-view
    ^
Error: Unrecognized command found at '^' position.
<R1>quit
```

可以看到，用户 user1@huawei Telnet R1 后不能使用命令 system-view 进入系统视图，原因是用户操作权限配置的是级别 0，因此操作受限。

步骤三 在 R3 上配置 AAA 功能

（1）在 R3 上配置认证方案为本地认证，授权方案为本地授权。

```
[R3]aaa
[R3-aaa]authentication-scheme auth1
Info: Create a new authentication scheme.
[R3-aaa-authen-auth1]authentication-mode local
[R3-aaa-authen-auth1]quit
[R3-aaa]authorization-scheme auth2
Info: Create a new authorization scheme.
[R3-aaa-author-auth2]authorization-mode local
[R3-aaa-author-auth2]quit
```

（2）在 R3 上创建域"huawei"并将认证方案和授权方案与域关联起来，然后创建一个用户并将用户加入域"huawei"。

```
[R3]telnet server enable
[R3]aaa
[R3-aaa]domain huawei
[R3-aaa-domain-huawei]authentication-scheme auth1
[R3-aaa-domain-huawei]authorization-scheme auth2
[R3-aaa-domain-huawei]quit
[R3-aaa]local-user user3@huawei password cipher huawei123
[R3-aaa]local-user user3@huawei service-type telnet
[R3-aaa]local-user user3@huawei privilege level 0
```

（3）将 R3 配置为 Telnet 服务器，并将认证模式配置为 AAA。

```
[R3]user-interface vty 0 4
[R3-ui-vty0-4]authentication-mode aaa
```

（4）验证 R1 Telnet R3 时是否要经过 AAA 认证。

```
<R1>telnet 119.84.111.3
  Press CTRL_] to quit telnet mode
  Trying 119.84.111.1 ...
  Connected to 119.84.111.1 ...
Login authentication
Username:user3@huawei
Password:
<R3>system-view
      ^
Error: Unrecognized command found at '^' position.
<R3>
```

可以看到，用户 user3@huawei 同样是因为登录后操作权限配置的是级别 0，所以操作受限。

任务验证

验证 AAA 的配置结果：

```
<R1>display domain name huawei
  Domain-name                  : huawei
  Domain-state                 : Active
  Authentication-scheme-name   : auth1
  Accounting-scheme-name       : default
  Authorization-scheme-name    : auth2
  Service-scheme-name          : -
  RADIUS-server-template       : -
  HWTACACS-server-template     : -
  User-group                   : -
```

```
<R1>display local-user username user1@huawei
The contents of local user(s):
Password              : ****************
State                 : active
Service-type-mask     : T
Privilege level       : 0
Ftp-directory         : -
Access-limit          : -
Accessed-num          : 0
Idle-timeout          : -
User-group            : -
```

```
<R3>display local-user username user3@huawei
The contents of local user(s):
Password              : ****************
State                 : active
Service-type-mask     : T
Privilege level       : 0
Ftp-directory         : -
Access-limit          : -
Accessed-num          : 0
Idle-timeout          : -
User-group            : -
```

任务 3.2　IPSec VPN 配置

任务背景

企业对网络安全性的需求日益提升，而传统的 TCP/IP 协议缺乏有效的安全认证和保密机制。IPSec（Internet Protocol Security）作为一种开放标准的安全框架结构，可以用来保证 IP 数据报文在网络上传输的机密性、完整性和防重放。

企业远程分支机构可以通过使用 IPSec VPN 建立安全传输通道，接入企业总部网络，如图 8-36 所示。

IPSec VPN 配置

图 8-36　IPSec 隧道

任务规划

企业的某些私有数据在公网传输时要确保完整性和机密性。作为企业的网络管理员，您需要在企业总部的边缘路由器（R1）和分支机构路由器（R3）之间部署 IPSec VPN 解决方案，建立 IPSec 隧道，用于安全传输来自指定部门的数据流。IPSec VPN 实验拓扑图如图 8-37 所示。

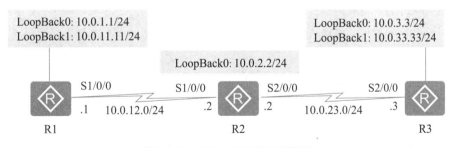

图 8-37　IPSec VPN 实验拓扑图

任务实施

步骤一　实验环境准备

```
<Huawei>system-view
[Huawei]sysname R1
[R1]interface Serial 1/0/0
[R1-Serial1/0/0]ip address 10.0.12.1 24
```

```
[R1-Serial1/0/0]quit
[R1]interface loopback 0
[R1-LoopBack0]ip address 10.0.1.1 24
```

```
<Huawei>system-view
[Huawei]sysname R2
[R2]interface Serial 1/0/0
[R2-Serial1/0/0]ip address 10.0.12.2 24
[R2-Serial1/0/0]quit
[R2]interface serial 2/0/0
[R2-Serial2/0/0]ip address 10.0.23.2 24
[R2-Serial2/0/0]quit
[R2]interface loopback 0
[R2-LoopBack0]ip address 10.0.2.2 24
```

```
<Huawei>system-view
[Huawei]sysname R3
[R3]interface Serial 2/0/0
[R3-Serial2/0/0]ip address 10.0.23.3 24
[R3-Serial2/0/0]quit
[R3]interface loopback 0
[R3-LoopBack0]ip address 10.0.3.3 24
```

步骤二　创建逻辑接口

```
[R1-LoopBack0]interface loopback 1
[R1-LoopBack1]ip address 10.0.11.11 24
```

```
[R3-LoopBack0]interface loopback 1
[R3-LoopBack1]ip address 10.0.33.33 24
```

步骤三　配置 OSPF

在 R1、R2 和 R3 上配置 OSPF，将 LoopBack0 的 IP 地址作为路由器的 ID，使用 OSPF 的默认进程 1，并将公网网段 10.0.12.0/24 和 10.0.23.0/24 以及环回接口地址通告在 OSPF 区域 0。

```
[R1]ospf router-id 10.0.1.1
[R1-ospf-1]area 0
[R1-ospf-1-area-0.0.0.0]network 10.0.12.0 0.0.0.255
[R1-ospf-1-area-0.0.0.0]network 10.0.1.0 0.0.0.255
[R1-ospf-1-area-0.0.0.0]network 10.0.11.0 0.0.0.255
```

```
[R2]ospf router-id  10.0.2.2
[R2-ospf-1]area 0
[R2-ospf-1-area-0.0.0.0]network  10.0.2.0 0.0.0.255
[R2-ospf-1-area-0.0.0.0]network  10.0.12.0 0.0.0.255
[R2-ospf-1-area-0.0.0.0]network  10.0.23.0 0.0.0.255

[R3]ospf router-id  10.0.3.3
[R3-ospf-1]area 0
[R3-ospf-1-area-0.0.0.0]network  10.0.23.0 0.0.0.255
[R3-ospf-1-area-0.0.0.0]network  10.0.3.0 0.0.0.255
[R3-ospf-1-area-0.0.0.0]network  10.0.33.0 0.0.0.255
```

待 OSPF 收敛完成后，查看 OSPF 邻居以及路由表。

```
<R2>display ospf peer brief
         OSPF Process 1 with Router ID 10.0.2.2
              Peer Statistic Information
-----------------------------------------------------------------
Area Id         Interface                 Neighbor id        State
0.0.0.0         Serial1/0/0               10.0.1.1           Full
0.0.0.0         Serial2/0/0               10.0.3.3           Full
-----------------------------------------------------------------


<R1>display ip routing-table
Route Flags: R - relay, D - download to fib
-----------------------------------------------------------------
Routing Tables: Public
         Destinations : 17     Routes : 17
Destination/Mask     Proto   Pre   Cost  Flags  NextHop        Interface
10.0.1.0/24          Direct  0     0     D      10.0.1.1       LoopBack0
10.0.1.1/32          Direct  0     0     D      127.0.0.1      LoopBack0
10.0.1.255/32        Direct  0     0     D      127.0.0.1      LoopBack0
10.0.2.2/32          OSPF    10    781   D      10.0.12.2      Serial1/0/0
10.0.3.3/32          OSPF    10    2343  D      10.0.12.2      Serial1/0/0
10.0.11.0/24         Direct  0     0     D      10.0.11.11     LoopBack1
10.0.11.11/32        Direct  0     0     D      127.0.0.1      LoopBack1
10.0.11.255/32       Direct  0     0     D      127.0.0.1      LoopBack1
10.0.12.0/24         Direct  0     0     D      10.0.12.1      Serial1/0/0
10.0.12.1/32         Direct  0     0     D      127.0.0.1      Serial1/0/0
10.0.12.2/32         Direct  0     0     D      10.0.12.2      Serial1/0/0
10.0.12.255/32       Direct  0     0     D      127.0.0.1      Serial1/0/0
```

```
10.0.23.0/24         OSPF    10   2343   D   10.0.12.2   Serial1/0/0
10.0.33.33/32        OSPF    10   2343   D   10.0.12.2   Serial1/0/0
127.0.0.0/8          Direct  0    0      D   127.0.0.1   InLoopBack0
127.0.0.1/32         Direct  0    0      D   127.0.0.1   InLoopBack0
127.255.255.255/32   Direct  0    0      D   127.0.0.1   InLoopBack0
255.255.255.255/32   Direct  0    0      D   127.0.0.1   InLoopBack0
```

步骤四 配置 ACL 定义感兴趣流量

配置高级 ACL 来定义 IPSec VPN 的感兴趣流量。高级 ACL 能够基于特定的参数来匹配流量。

```
[R1]acl 3001
[R1-acl-adv-3001]rule 5 permit ip source 10.0.1.0 0.0.0.255 destination 10.0.3.0 0.0.0.255


[R3]acl 3001
[R3-acl-adv-3001]rule 5 permit ip source 10.0.3.0 0.0.0.255 destination 10.0.1.0 0.0.0.255
```

步骤五 配置 IPSec VPN 提议

创建 IPSec 提议，并进入 IPSec 提议视图来指定安全协议。注意确保隧道两端的设备使用相同的安全协议。

```
[R1]ipsec proposal tran1
[R1-ipsec-proposal-tran1]esp authentication-algorithm sha1
[R1-ipsec-proposal-tran1]esp encryption-algorithm 3des


[R3]ipsec proposal tran1
[R3-ipsec-proposal-tran1]esp authentication-algorithm sha1
[R3-ipsec-proposal-tran1]esp encryption-algorithm 3des
```

执行 display ipsec proposal 命令，验证配置结果。

```
[R1]display ipsec proposal
Number of proposals: 1
IPSec proposal name       : tran1
  Encapsulation mode      : Tunnel
  Transform               : esp-new
  ESP protocol       :    Authentication SHA1-HMAC-96
Encryption      3DES
```

```
[R3]display ipsec proposal
Number of proposals: 1
IPSec proposal name    : tran1
  Encapsulation mode   : Tunnel
  Transform            : esp-new
  ESP protocol         :    Authentication SHA1-HMAC-96
Encryption     3DES
```

步骤六 创建 IPSec 策略

手工创建 IPSec 策略，每一个 IPSec 安全策略都使用唯一的名称和序号来标识，IPSec 策略中会应用 IPSec 提议中定义的安全协议、认证算法、加密算法和封装模式，手工创建的 IPSec 策略还需要配置安全联盟（SA）中的参数。

```
[R1]ipsec policy P1 10 manual
[R1-ipsec-policy-manual-P1-10]security acl 3001
[R1-ipsec-policy-manual-P1-10]proposal tran1
[R1-ipsec-policy-manual-P1-10]tunnel remote 10.0.23.3
[R1-ipsec-policy-manual-P1-10]tunnel local 10.0.12.1
[R1-ipsec-policy-manual-P1-10]sa spi outbound esp 54321
[R1-ipsec-policy-manual-P1-10]sa spi inbound esp 12345
[R1-ipsec-policy-manual-P1-10]sa string-key outbound esp simple huawei
[R1-ipsec-policy-manual-P1-10]sa string-key inbound esp simple huawei

[R3]ipsec policy P1 10 manual
[R3-ipsec-policy-manual-P1-10]security acl 3001
[R3-ipsec-policy-manual-P1-10]proposal tran1
[R3-ipsec-policy-manual-P1-10]tunnel remote 10.0.12.1
[R3-ipsec-policy-manual-P1-10]tunnel local 10.0.23.3
[R3-ipsec-policy-manual-P1-10]sa spi outbound esp 12345
[R3-ipsec-policy-manual-P1-10]sa spi inbound esp 54321
[R3-ipsec-policy-manual-P1-10]sa string-key outbound esp simple huawei
[R3-ipsec-policy-manual-P1-10]sa string-key inbound esp simple huawei
```

执行 display ipsec policy 命令，验证配置结果。

```
<R1>display ipsec policy
===========================================
IPSec policy group: "P1"
Using interface:
===========================================
    Sequence number: 10
    Security data flow: 3001
```

```
        Tunnel local   address: 10.0.12.1
        Tunnel remote address: 10.0.23.3
        Qos pre-classify: Disable
        Proposal name:tran1
        Inbound AH setting:
            AH SPI:
            AH string-key:
            AH authentication hex key:
        Inbound ESP setting:
            ESP SPI: 12345 (0x3039)
            ESP string-key: huawei
            ESP encryption hex key:
            ESP authentication hex key:
        Outbound AH setting:
            AH SPI:
            AH string-key:
            AH authentication hex key:
        Outbound ESP setting:
            ESP SPI: 54321 (0xd431)
            ESP string-key: huawei
            ESP encryption hex key:
            ESP authentication hex key:
```

步骤七 在接口下应用 IPSec 策略

在物理接口应用 IPSec 策略，接口将对感兴趣流量进行 IPSec 加密处理。

```
[R1]interface Serial 1/0/0
[R1-Serial1/0/0]ipsec policy P1
```

```
[R3]interface Serial 2/0/0
[R3-Serial2/0/0]ipsec policy P1
```

任务验证

检测网络的连通性：
（1）验证设备对不感兴趣流量不进行 IPSec 加密处理。

```
<R1>ping -a 10.0.11.11 10.0.33.33
   PING 10.0.33.33: 56  data bytes, press CTRL_C to break
     Reply from 10.0.33.33: bytes=56 Sequence=1 ttl=254 time=60 ms
     Reply from 10.0.33.33: bytes=56 Sequence=2 ttl=254 time=50 ms
     Reply from 10.0.33.33: bytes=56 Sequence=3 ttl=254 time=50 ms
```

```
  Reply from 10.0.33.33: bytes=56 Sequence=4 ttl=254 time=60 ms
  Reply from 10.0.33.33: bytes=56 Sequence=5 ttl=254 time=50 ms
---10.0.33.33 ping statistics ---
  5 packet(s) transmitted
  5 packet(s) received
  0.00%packet loss
  round-trip min/avg/max =50/54/60 ms

<R1>display ipsec statistics esp
  Inpacket count                     : 0
  Inpacket auth count                : 0
  Inpacket decap count               : 0
  Outpacket count                    : 0
  Outpacket auth count               : 0
  Outpacket encap count              : 0
  Inpacket drop count                : 0
  Outpacket drop count               : 0
  BadAuthLen count                   : 0
  AuthFail count                     : 0
  InSAAclCheckFail count             : 0
  PktDuplicateDrop count             : 0
  PktSeqNoTooSmallDrop count         : 0
  PktInSAMissDrop count              : 0
```

（2）验证设备将对感兴趣流量进行 IPSec 加密处理。

```
<R1>ping -a 10.0.1.1 10.0.3.3
  PING 10.0.3.3: 56   data bytes, press CTRL_C to break
    Reply from 10.0.3.3: bytes=56 Sequence=1 ttl=255 time=80 ms
    Reply from 10.0.3.3: bytes=56 Sequence=2 ttl=255 time=77 ms
    Reply from 10.0.3.3: bytes=56 Sequence=3 ttl=255 time=77 ms
    Reply from 10.0.3.3: bytes=56 Sequence=4 ttl=255 time=80 ms
    Reply from 10.0.3.3: bytes=56 Sequence=5 ttl=255 time=77 ms
---10.0.3.3 ping statistics ---
  5 packet(s) transmitted
  5 packet(s) received
  0.00%packet loss
  round-trip min/avg/max =77/78/80 ms

<R1>display ipsec statistics esp
  Inpacket count                     : 5
  Inpacket auth count                : 0
  Inpacket decap count               : 0
```

```
Outpacket count              : 5
Outpacket auth count         : 0
Outpacket encap count        : 0
Inpacket drop count          : 0
Outpacket drop count         : 0
BadAuthLen count             : 0
AuthFail count               : 0
InSAAclCheckFail count       : 0
PktDuplicateDrop count       : 0
PktSeqNoTooSmallDrop count   : 0
PktInSAMissDrop count        : 0
```

任务 3.3　GRE 隧道配置

📖 任务背景

GRE 隧道配置

　　IPSec VPN 用于在两个端点之间提供安全的 IP 通信，但只能加密并传播单播数据，无法加密和传输语音、视频、动态路由协议信息等组播数据流量。

　　通用路由封装（Generic Routing Encapsulation，GRE）协议提供了将一种协议的报文封装在另一种协议的报文中的机制，是一种隧道封装技术。GRE 可以封装组播数据，并可以和 IPSec 结合使用，从而保证语音、视频等组播业务的安全。

　　IPSec VPN 技术可以创建一条跨越共享公网的隧道，从而实现私网互联。IPSec VPN 能够安全传输 IP 报文，但是无法在隧道的两个端点之间运行 RIP 和 OSPF 等路由协议。GRE 可以将路由协议信息封装在另一种协议报文（如 IP）中进行传输。

　　GRE 支持将一种协议的报文封装在另一种协议报文中，可以解决异种网络的传输问题。GRE 隧道扩展了受跳数限制的路由协议的工作范围，支持企业灵活设计网络拓扑，如图 8-38 所示。

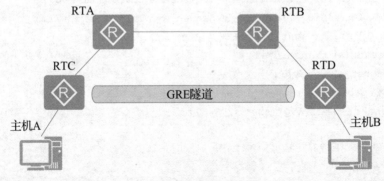

图 8-38　GRE 隧道应用

任务规划

您是企业的网络管理员,当企业总部和分支机构间需要互相发布加密的路由信息时,仅通过 IPSec VPN 方案是无法实现的。由于 IPSec VPN 无法承载使用组播发送的路由协议数据包,因此您还需要在现有的 IPSec 网络中配置 GRE 隧道解决此问题。GRE 隧道配置实验拓扑图如图 8-39 所示。

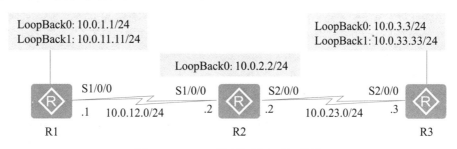

图 8-39 GRE 隧道配置实验拓扑图

任务实施

步骤一 创建 GRE 隧道

创建隧道接口并为该接口配置一个公网 IP 地址,然后指定接口封装类型为 GRE,并配置隧道的实际源地址以及实际目的地址。

```
[R1]interface Tunnel 0/0/1
[R1-Tunnel0/0/1]ip address 100.1.1.1 24
[R1-Tunnel0/0/1]tunnel-protocol gre
[R1-Tunnel0/0/1]source 10.0.12.1
[R1-Tunnel0/0/1]destination 10.0.23.3
```

```
[R3]interface Tunnel 0/0/1
[R3-Tunnel0/0/1]ip address 100.1.1.2 24
[R3-Tunnel0/0/1]tunnel-protocol gre
[R3-Tunnel0/0/1]source 10.0.23.3
[R3-Tunnel0/0/1]destination 10.0.12.1
```

步骤二 配置 OSPF 进程 2 用于隧道路由

将隧道接口所在的网络通告在 OSPF 进程 1,从 OSPF 进程 1 中删除网络 10.0.12.0/24 和 10.0.23.0/24。创建链 OSPF 进程 2,并将网络 10.0.12.0/24 和 10.0.23.0/24 通告到 OSPF 进程 2。

```
[R1]ospf 1
[R1-ospf-1]area 0
[R1-ospf-1-area-0.0.0.0]network 100.1.1.0 0.0.0.255
[R1-ospf-1-area-0.0.0.0]undo network 10.0.12.0 0.0.0.255
```

```
[R1]ospf 2 router-id 10.0.1.1
[R1-ospf-2]area 0
[R1-ospf-2-area-0.0.0.0]network 10.0.12.0 0.0.0.255
```

```
[R3]ospf 1
[R3-ospf-1]area 0
[R3-ospf-1-area-0.0.0.0]network 100.1.1.0 0.0.0.255
[R3-ospf-1-area-0.0.0.0]undo network 10.0.23.0 0.0.0.255
[R3]ospf 2 router-id 10.0.3.3
[R3-ospf-2]area 0
[R3-ospf-2-area-0.0.0.0]network 10.0.23.0 0.0.0.255
```

路由器会为不同的 OSPF 进程创建不同的 LSDB，R1 和 R3 中分别有 LSDB 1 和 LSDB 2，两个数据库彼此独立，不会同步路由信息。因此 R2 学习不到 R1 和 R3 通告在进程 2 中的路由。

执行 display interface Tunnel 0/0/1 命令，验证配置结果。

```
<R1>display interface Tunnel 0/0/1
Tunnel0/0/1 current state : UP
Line protocol current state : UP
Last line protocol up time : 2016-03-17 17:10:16
Description:HUAWEI, AR Series, Tunnel0/0/1 Interface
Route Port,The Maximum Transmit Unit is 1500
Internet Address is 100.1.1.1/24
Encapsulation is TUNNEL, loopback not set
Tunnel source 10.0.12.1 (Serial1/0/0), destination 10.0.23.3
Tunnel protocol/transport GRE/IP, key disabled
keepalive disabled
Checksumming of packets disabled
Current system time: 2016-03-17 17:35:39
    Last 300 seconds input rate 0 bytes/sec, 0 packets/sec
    Last 300 seconds output rate 9 bytes/sec, 0 packets/sec
    Realtime 0 seconds input rate 0 bytes/sec, 0 packets/sec
    Realtime 0 seconds output rate 0 bytes/sec, 0 packets/sec
    0 packets input, 0 bytes, 0 drops
    145 packets output, 14320 bytes, 0 drops
    Input bandwidth utilization  : --
    Output bandwidth utilization : --
```

步骤三 将 GRE 流量定义为感兴趣流量

重新配置 ACL 定义感兴趣流量。

```
[R1]acl 3001
[R1-acl-adv-3001]rule 5 permit gre source 10.0.12.1 0 destination 10.0.23.3 0

[R3]acl 3001
[R3-acl-adv-3001]rule 5 permit gre source 10.0.23.3 0 destination 10.0.12.1 0
```

步骤四 给 GRE 隧道配置 Keepalive 功能

```
[R1]interface Tunnel 0/0/1
[R1-Tunnel0/0/1]keepalive period 3
```

任务验证

步骤一 验证路由信息通过 GRE 封装后可由 IPSec VPN 传输

执行 display ip routing-table 命令，查看 IPv4 路由表。

```
<R1>display ip routing-table
Route Flags: R - relay, D - download to fib
------------------------------------------------------------------------
Routing Tables: Public
         Destinations : 21         Routes : 21
Destination/Mask      Proto   Pre  Cost  Flags  NextHop      Interface
10.0.1.0/24           Direct  0    0     D      10.0.1.1     LoopBack0
10.0.1.1/32           Direct  0    0     D      127.0.0.1    LoopBack0
10.0.1.255/32         Direct  0    0     D      127.0.0.1    LoopBack0
10.0.2.2/32           OSPF    10   781   D      10.0.12.2    Serial1/0/0
10.0.3.3/32           OSPF    10   1562  D      100.1.1.2    Tunnel0/0/1
10.0.11.0/24          Direct  0    0     D      10.0.11.11   LoopBack1
10.0.11.11/32         Direct  0    0     D      127.0.0.1    LoopBack1
10.0.11.255/32        Direct  0    0     D      127.0.0.1    LoopBack1
10.0.12.0/24          Direct  0    0     D      10.0.12.1    Serial1/0/0
10.0.12.1/32          Direct  0    0     D      127.0.0.1    Serial1/0/0
10.0.12.2/32          Direct  0    0     D      10.0.12.2    Serial1/0/0
10.0.12.255/32        Direct  0    0     D      127.0.0.1    Serial1/0/0
10.0.23.0/24          OSPF    10   2343  D      10.0.12.2    Serial1/0/0
10.0.33.33/32         OSPF    10   1562  D      100.1.1.2    Tunnel0/0/1
100.1.1.0/24          Direct  0    0     D      100.1.1.1    Tunnel0/0/1
100.1.1.1/32          Direct  0    0     D      127.0.0.1    Tunnel0/0/1
100.1.1.255/32        Direct  0    0     D      127.0.0.1    Tunnel0/0/1
127.0.0.0/8           Direct  0    0     D      127.0.0.1    InLoopBack0
127.0.0.1/32          Direct  0    0     D      127.0.0.1    InLoopBack0
127.255.255.255/32    Direct  0    0     D      127.0.0.1    InLoopBack0
255.255.255.255/32    Direct  0    0     D      127.0.0.1    InLoopBack0
```

可以观察到，GRE 隧道建立后，路由器可以将 OSPF 协议报文通过 GRE 封装后进行交互，从而获取对端路由信息。

步骤二 清除 IPSec 统计信息后，再通过 ping 命令测试网络连通性

```
<R1>reset ipsec statistics esp
[R1]ping -a 10.0.1.1 10.0.3.3
  PING 10.0.3.3: 56  data bytes, press CTRL_C to break
    Reply from 10.0.3.3: bytes=56 Sequence=1 ttl=255 time=69 ms
    Reply from 10.0.3.3: bytes=56 Sequence=2 ttl=255 time=70 ms
    Reply from 10.0.3.3: bytes=56 Sequence=3 ttl=255 time=68 ms
    Reply from 10.0.3.3: bytes=56 Sequence=4 ttl=255 time=68 ms
    Reply from 10.0.3.3: bytes=56 Sequence=5 ttl=255 time=68 ms
  ---10.0.3.3 ping statistics ---
    5 packet(s) transmitted
    5 packet(s) received
    0.00%packet loss
round-trip min/avg/max =68/68/70 ms

<R1>display ipsec statistics esp
  Inpacket count              : 8
  Inpacket auth count         : 0
  Inpacket decap count        : 0
  Outpacket count             : 8
  Outpacket auth count        : 0
  Outpacket encap count       : 0
  Inpacket drop count         : 0
  Outpacket drop count        : 0
  BadAuthLen count            : 0
  AuthFail count              : 0
  InSAAclCheckFail count      : 0
  PktDuplicateDrop count      : 0
  PktSeqNoTooSmallDrop count  : 0
  PktInSAMissDrop count       : 0
```

从上述 IPSec ESP 统计信息可以看出，OSPF 协议交互的报文（包括 hello 报文）进行了 GRE 封装后再被 IPSec VPN 加密传输。

步骤三 验证隧道接口的 Keepalive 功能是否已开启

```
<R1>display interface Tunnel 0/0/1
Tunnel0/0/1 current state : UP
Line protocol current state : UP
```

```
Last line protocol up time : 2016-03-18 09:50:21
Description:HUAWEI, AR Series, Tunnel0/0/1 Interface
Route Port,The Maximum Transmit Unit is 1500
Internet Address is 100.1.1.1/24
Encapsulation is TUNNEL, loopback not set
Tunnel source 10.0.12.1 (Serial1/0/0), destination 10.0.23.3
Tunnel protocol/transport GRE/IP, key disabled
keepalive enable period 3 retry-times 3
Checksumming of packets disabled
Current system time: 2016-03-18 11:05:49
    Last 300 seconds input rate 0 bytes/sec, 0 packets/sec
    Last 300 seconds output rate 8 bytes/sec, 0 packets/sec
    Realtime 0 seconds input rate 0 bytes/sec, 0 packets/sec
    Realtime 0 seconds output rate 0 bytes/sec, 0 packets/sec
    0 packets input, 0 bytes, 0 drops
    503 packets output, 47444 bytes, 0 drops
    Input bandwidth utilization  : --
    Output bandwidth utilization : --
```

项目小结

网络安全问题与日俱增，网络安全技术愈发重要。通过本项目的学习，同学们掌握了通过 ACL 控制公司内、外用户对网络的访问的方法，学会了使用 NAT 技术解决公司 IP 地址紧缺的情况，学习了如何利用 AAA 管理用户、分配访问权限等，并且了解了 IPSec 和 GRE 两种隧道传输技术，使内部数据安全地在网络上传输，保障数据的安全性。

项目 9 IPv6 协议

项目目标

1. 掌握基于 IPv6 的静态路由的配置方法
2. 掌握基于 IPv6 的默认路由的配置方法

任务 1 基于 IPv6 的静态路由

基于 IPv6 的静态路由

任务背景

某企业网络需要在网络内部署 IPv6 协议并实现 IPv6 的互联互通，需要对当前运行的网络设备进行配置。

任务规划

某公司有北京总部、上海分部和广州分部 3 个办公地点，各分部与总部之间使用路由器互联。北京、上海、广州的路由器分别为 R1、R2、R3，全网使用 IPv6 进行组网，需要配置静态路由，使所有计算机能够互相访问。网络拓扑图如图 9-1 所示，IP 地址规划见表 9-1。

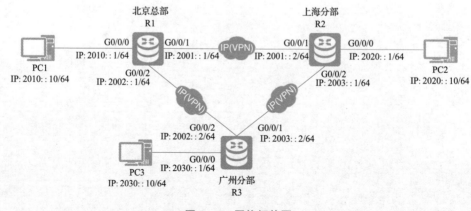

图 9-1 网络拓扑图

表 9-1 IP 地址规划表

设备	接口	IP 地址
R1	G0/0/0	2010::1/64
R1	G0/0/1	2001::1/64
R1	G0/0/2	2002::1/64
R2	G0/0/0	2020::1/64
R2	G0/0/1	2001::2/64
R2	G0/0/2	2003::1/64
R3	G0/0/0	2030::1/64
R3	G0/0/1	2003::2/64
R3	G0/0/2	2002::2/64
PC1	E0/0/1	2010::10/64
PC2	E0/0/1	2020::10/64
PC3	E0/0/1	2030::10/64

配置步骤如下：

（1）配置设备及接口 IPv6 功能；

（2）配置静态路由；

（3）配置各计算机的 IP 地址。

任务实施

步骤一　配置设备及接口 IPv6 功能

IPv6 命令用来使能设备转发 IPv6 单播报文，包括本地 IPv6 报文的发送与接收。

（1）R1 的配置。

```
[Huawei]system-view
[Huawei]sysname R1
[R1]ipv6
[R1]interface GigabitEthernet 0/0/0
[R1-GigabitEthernet0/0/0]ipv6 enable
[R1-GigabitEthernet0/0/0]ipv6 address 2010::1/64
[R1]interface GigabitEthernet 0/0/1
[R1-GigabitEthernet0/0/1]ipv6 enable
[R1-GigabitEthernet0/0/1]ipv6 address 2001::1/64
[R1]interface GigabitEthernet 0/0/2
[R1-GigabitEthernet0/0/2]ipv6 enable
[R1-GigabitEthernet0/0/2]ipv6 address 2002::1/64
```

（2）R2 的配置。

```
[Huawei]system-view
[Huawei]sysname R2
[R2]ipv6
[R2]interface GigabitEthernet 0/0/0
[R2-GigabitEthernet0/0/0]ipv6 enable
[R2-GigabitEthernet0/0/0]ipv6 address 2020::1/64
[R2]interface GigabitEthernet 0/0/1
[R2-GigabitEthernet0/0/1]ipv6 enable
[R2-GigabitEthernet0/0/1]ipv6 address 2001::2/64
[R2]interface GigabitEthernet 0/0/2
[R2-GigabitEthernet0/0/2]ipv6 enable
[R2-GigabitEthernet0/0/2]ipv6 address 2003::1/64
```

（3）R3 的配置。

```
[Huawei]system-view
[Huawei]sysname R3
[R3]ipv6
[R3]interface GigabitEthernet 0/0/0
[R3-GigabitEthernet0/0/0]ipv6 enable
[R3-GigabitEthernet0/0/0]ipv6 address 2030::1/64
[R3]interface GigabitEthernet 0/0/1
[R3-GigabitEthernet0/0/1]ipv6 enable
[R3-GigabitEthernet0/0/1]ipv6 address 2003::2/64
[R3]interface GigabitEthernet 0/0/2
[R3-GigabitEthernet0/0/2]ipv6 enable
[R3-GigabitEthernet0/0/2]ipv6 address 2002::2/64
```

步骤二 配置静态路由

在 R1 上配置静态路由。

```
[R1]ipv6 route-static 2020:: 64 2001::2
[R1]ipv6 route-static 2030:: 64 2002::2
```

在 R2 上配置静态路由。

```
[R2]ipv6 route-static 2010:: 64 2001::1
[R2]ipv6 route-static 2030:: 64 2003::2
```

在 R3 上配置静态路由。

```
[R3]ipv6 route-static 2010:: 64 2002::1
[R3]ipv6 route-static 2020:: 64 2003::1
```

步骤三 配置各计算机的 IP 地址

根据拓扑图中的地址配置各主机的 IP 地址，如图 9-2～图 9-4 所示。

图 9-2　PC1 IP 地址配置图

图 9-3　PC2 IP 地址配置图

路由交换技术

图 9-4　PC3 IP 地址配置图

任务验证

步骤一　验证路由器上端口的配置信息

在路由器上使用 display ipv6 interface brief 命令，查看配置信息。

（1）R1 的配置。

```
[R1]display ipv6 interface brief
*down: administratively down
(l): loopback
(s): spoofing
Interface                   Physical           Protocol
GigabitEthernet0/0/0        up                 up
[IPv6 Address] 2010::1
GigabitEthernet0/0/1        up                 up
[IPv6 Address] 2001::1
GigabitEthernet0/0/2        up                 up
[IPv6 Address] 2002::1
```

（2）R2 的配置。

```
[R2]display ipv6 interface brief
*down: administratively down
(l): loopback
(s): spoofing
```

```
Interface                     Physical           Protocol
GigabitEthernet0/0/0          up                 up
 [IPv6 Address] 2020::1
GigabitEthernet0/0/1          up                 up
 [IPv6 Address] 2001::2
GigabitEthernet0/0/2          up                 up
 [IPv6 Address] 2003::1
```

(3) R3 的配置。

```
[R3]display ipv6 interface brief
*down: administratively down
(l): loopback
(s): spoofing
Interface                     Physical           Protocol
GigabitEthernet0/0/0          up                 up
 [IPv6 Address] 2030::1
GigabitEthernet0/0/1          up                 up
 [IPv6 Address] 2003::2
GigabitEthernet0/0/2          up                 up
 [IPv6 Address] 2002::2
```

步骤二　查看路由表的信息

在路由器上使用 display ipv6 routing-table 命令，查看路由表信息。

(1) R1 的配置。

```
[R1]display ipv6 routing-table
Routing Table : Public
Destinations : 10     Routes : 10

Destination  : ::1                    PrefixLength : 128
NextHop      : ::1                    Preference   : 0
Cost         : 0                      Protocol     : Direct
RelayNextHop : ::                     TunnelID     : 0x0
Interface    : InLoopBack0            Flags        : D

Destination  : 2001::                 PrefixLength : 64
NextHop      : 2001::1                Preference   : 0
Cost         : 0                      Protocol     : Direct
RelayNextHop : ::                     TunnelID     : 0x0
Interface    : GigabitEthernet0/0/1   Flags        : D
```

```
Destination  : 2001::1                    PrefixLength : 128
NextHop      : ::1                        Preference   : 0
Cost         : 0                          Protocol     : Direct
RelayNextHop : ::                         TunnelID     : 0x0
Interface    : GigabitEthernet0/0/1       Flags        : D

Destination  : 2002::                     PrefixLength : 64
NextHop      : 2002::1                    Preference   : 0
Cost         : 0                          Protocol     : Direct
RelayNextHop : ::                         TunnelID     : 0x0
Interface    : GigabitEthernet0/0/2       Flags        : D

Destination  : 2002::1                    PrefixLength : 128
NextHop      : ::1                        Preference   : 0
Cost         : 0                          Protocol     : Direct
RelayNextHop : ::                         TunnelID     : 0x0
Interface    : GigabitEthernet0/0/2       Flags        : D

Destination  : 2010::                     PrefixLength : 64
NextHop      : 2010::1                    Preference   : 0
Cost         : 0                          Protocol     : Direct
RelayNextHop : ::                         TunnelID     : 0x0
Interface    : GigabitEthernet0/0/0       Flags        : D

Destination  : 2010::1                    PrefixLength : 128
NextHop      : ::1                        Preference   : 0
Cost         : 0                          Protocol     : Direct
RelayNextHop : ::                         TunnelID     : 0x0
Interface    : GigabitEthernet0/0/0       Flags        : D

Destination  : 2020::                     PrefixLength : 64
NextHop      : 2001::2                    Preference   : 60
Cost         : 0                          Protocol     : Static
RelayNextHop : ::                         TunnelID     : 0x0
Interface    : GigabitEthernet0/0/1       Flags        : RD

Destination  : 2030::                     PrefixLength : 64
NextHop      : 2002::2                    Preference   : 60
Cost         : 0                          Protocol     : Static
RelayNextHop : ::                         TunnelID     : 0x0
Interface    : GigabitEthernet0/0/2       Flags        : RD
```

```
Destination : FE80::              PrefixLength : 10
NextHop     : ::                  Preference   : 0
Cost        : 0                   Protocol     : Direct
RelayNextHop: ::                  TunnelID     : 0x0
Interface   : NULL0               Flags        : D
```

(2) 同理,可以查看 R2 和 R3 的路由表。

步骤三 测试计算机的连通性

通过 ping 命令,测试发现各计算机之间可以互相通信,如图 9-5 所示。

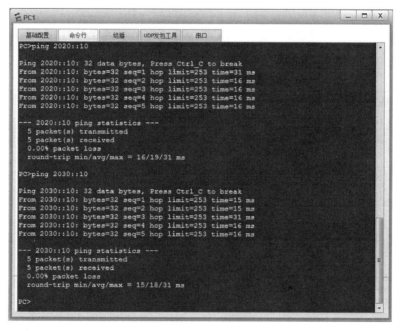

图 9-5 测试 PC 连通性 1

任务 2　基于 IPv6 的单臂路由

🔲 任务背景

某公司有北京总部、上海分部和广州分部 3 个办公地点,各分部与总部之间使用路由器互联。北京、上海、广州的路由器分别为 R1、R2、R3,全网使用 IPv6 进行组网。

基于 IPv6 的单臂路由

🔲 任务规划

某公司北京总部共有财务部、市场部和技术部 3 个部门,需划分独立的 VLAN。该公司现需要对路由器进行相应配置，使所有计算机能够互相访问。网络拓扑图如图 9-6 所示，IP 地址规划见表 9-2。

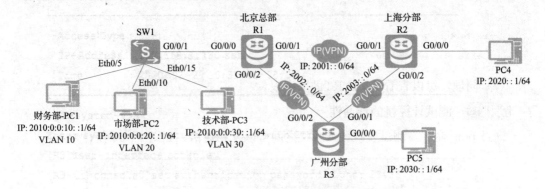

图 9-6　网络拓扑图

表 9-2　IP 地址规划表

设备	接口	IP 地址
R1	G0/0/0.10	2010:0:0:10::1/64
R1	G0/0/0.20	2010:0:0:20::1/64
R1	G0/0/0.30	2010:0:0:30::1/64
R1	G0/0/1	2001::1/64
R1	G0/0/2	2002::1/64
R2	G0/0/0	2020::1/64
R2	G0/0/1	2001::2/64
R2	G0/0/2	2003::2/64
R3	G0/0/0	2030::1/64
R3	G0/0/1	2003::3/64
R3	G0/0/2	2002::3/64
PC1	Eth0/0/1	2010:0:0:10::10
PC2	Eth0/0/1	2010:0:0:20::2
PC3	Eth0/0/1	2010:0:0:30::3
PC4	Eth0/0/1	2020::4
PC5	Eth0/0/1	2030::5

配置步骤如下：

（1）配置交换机接口；

（2）配置路由器；

（3）配置静态路由；

（4）配置各计算机的 IP 地址。

任务实施

步骤一　配置交换机接口

为各部门创建相应的 VLAN，将端口划分至相应的 VLAN。

```
[Huawei]system-view
[Huawei]sysname SW1
[SW1]vlan batch 10 20 30
[SW1]int Ethernet0/0/5
[SW1-Ethernet0/0/5]port link-type access
[SW1-Ethernet0/0/5]port default vlan 10
[SW1-Ethernet0/0/5]quit
[SW1]interface Ethernet0/0/10
[SW1-Ethernet0/0/10]port link-type access
[SW1-Ethernet0/0/10]port default vlan 20
[SW1-Ethernet0/0/10]quit
[SW1]interface Ethernet0/0/15
[SW1-Ethernet0/0/15]port link-type access
[SW1-Ethernet0/0/15]port default vlan 30
[SW1-Ethernet0/0/15]quit
[SW1]interface Ethernet0/0/1
[SW1-Ethernet0/0/1]port link-type trunk
[SW1]interface Ethernet0/0/1
[SW1-Ethernet0/0/1]port trunk allow-pass vlan 10 20 30
```

步骤二　配置路由器

在路由器 R1 以太网口上建立子接口，分别新建两个子端口，然后为两个子端口配置 IP 地址和掩码，作为 VLAN 的网关，同时启动 802.1Q 协议。

（1）R1 的配置。

```
[Huawei]system-view
[Huawei]sysname R1
[R1]ipv6
[R1]interface GigabitEthernet 0/0/0.10
[R1-GigabitEthernet0/0/0.10]dot1q termination vid 10
[R1-GigabitEthernet0/0/0.10]ipv6 enable
[R1-GigabitEthernet0/0/0.10]ipv6 address 2010:0:0:10::1/64
[R1-GigabitEthernet0/0/0.10]arp broadcast enable
[R1]interface GigabitEthernet 0/0/0.20
[R1-GigabitEthernet0/0/0.20]dot1q termination vid 20
[R1-GigabitEthernet0/0/0.20]ipv6 enable
[R1-GigabitEthernet0/0/0.20]ipv6 address 2010:0:0:20::1/64
[R1-GigabitEthernet0/0/0.20]arp broadcast enable
[R1]interface GigabitEthernet 0/0/0.30
[R1-GigabitEthernet0/0/0.30]dot1q termination vid 30
```

```
[R1-GigabitEthernet0/0/0.30]ipv6 enable
[R1-GigabitEthernet0/0/0.30]ipv6 address 2010:0:0:30::1/64
[R1-GigabitEthernet0/0/0.30]arp broadcast enable
[R1-GigabitEthernet0/0/0.30]quit
[R1]interface GigabitEthernet 0/0/1
[R1-GigabitEthernet0/0/1]ipv6 enable
[R1-GigabitEthernet0/0/1]ipv6 address 2001::1/64
[R1]interface GigabitEthernet 0/0/2
[R1-GigabitEthernet0/0/2]ipv6 enable
[R1-GigabitEthernet0/0/2]ipv6 address 2002::1/64
```

（2）R2 的配置。

```
[Huawei]system-view
[Huawei]sysname R2
[R2]interface GigabitEthernet 0/0/0
[R2-GigabitEthernet0/0/0]ipv6 enable
[R2-GigabitEthernet0/0/0]ipv6 address 2020::1/64
[R2]interface GigabitEthernet 0/0/1
[R2-GigabitEthernet0/0/1]ipv6 enable
[R2-GigabitEthernet0/0/1]ipv6 address 2001::2/64
[R2]interface GigabitEthernet 0/0/2
[R2-GigabitEthernet0/0/2]ipv6 enable
[R2-GigabitEthernet0/0/2]ipv6 address 2003::2/64
```

（3）R3 的配置。

```
[Huawei]system-view
[Huawei]sysname R3
[R3]interface GigabitEthernet 0/0/0
[R3-GigabitEthernet0/0/0]ipv6 enable
[R3-GigabitEthernet0/0/0]ipv6 address 2030::1/64
[R3]interface GigabitEthernet 0/0/1
[R3-GigabitEthernet0/0/1]ipv6 enable
[R3-GigabitEthernet0/0/1]ipv6 address 2003::3/64
[R3]interface GigabitEthernet 0/0/2
[R3-GigabitEthernet0/0/2]ipv6 enable
[R3-GigabitEthernet0/0/2]ipv6 address 2002::3/64
```

步骤三 配置静态路由

配置 R1 的路由。

```
[R1]ipv6 route-static 2020:: 64 2001::2
[R1]ipv6 route-static 2030:: 64 2002::3
```

配置 R2 的路由。

```
[R2]ipv6 route-static 2010:: 48 2001::1
[R2]ipv6 route-static 2030:: 64 2003::3
```

配置 R3 的路由。

```
[R3]ipv6 route-static 2010:: 48 2002::1
[R3]ipv6 route-static 2020:: 64 2003::2
```

步骤四 配置各计算机的 IP 地址

根据拓扑图中的地址配置各主机的 IP 地址，如图 9-7～图 9-11 所示。

图 9-7 财务部-PC1 IP 地址配置图

图 9-8 市场部-PC2 IP 地址配置图

图 9-9 技术部-PC3 IP 地址配置图

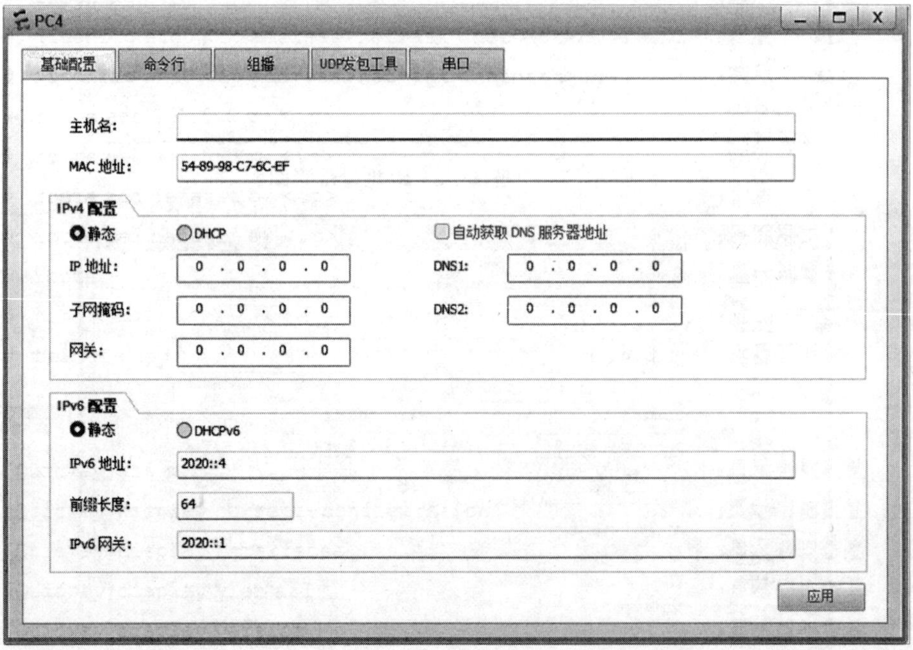

图 9-10 PC4 IP 地址配置图

项目 9 IPv6 协议

图 9-11　PC5 IP 地址配置图

任务验证

步骤一　验证交换机上端口的划分

在交换机上使用 display port vlan 命令，查看端口的划分。

```
[SW1]display port vlan
Port                Link Type    PVID    Trunk VLAN List
-------------------------------------------------------------------------
Ethernet0/0/1       hybrid       1       -
Ethernet0/0/2       hybrid       1       -
Ethernet0/0/3       hybrid       1       -
Ethernet0/0/4       hybrid       1       -
Ethernet0/0/5       access       10      -
Ethernet0/0/6       hybrid       1       -
Ethernet0/0/7       hybrid       1       -
Ethernet0/0/8       hybrid       1       -
Ethernet0/0/9       hybrid       1       -
Ethernet0/0/10      access       20      -
Ethernet0/0/11      hybrid       1       -
Ethernet0/0/12      hybrid       1       -
Ethernet0/0/13      hybrid       1       -
Ethernet0/0/14      hybrid       1       -
Ethernet0/0/15      access       30      -
Ethernet0/0/16      hybrid       1       -
```

```
Ethernet0/0/17          hybrid      1    -
Ethernet0/0/18          hybrid      1    -
Ethernet0/0/19          hybrid      1    -
Ethernet0/0/20          hybrid      1    -
Ethernet0/0/21          hybrid      1    -
Ethernet0/0/22          hybrid      1    -
GigabitEthernet0/0/1    trunk       1    1 10 20 30
GigabitEthernet0/0/2    hybrid      1    -
```

步骤二 验证路由器上端口的配置信息

在路由器上使用 display ipv6 interface brief 命令，查看配置信息。

（1）R1 的配置。

```
[R1]display ipv6 interface brief
*down: administratively down
(l): loopback
(s): spoofing
Interface                          Physical        Protocol
GigabitEthernet0/0/0.10            up              up
[IPv6 Address] 2010:0:0:10::1
GigabitEthernet0/0/0.20            up              up
[IPv6 Address] 2010:0:0:20::1
GigabitEthernet0/0/0.30            up              up
[IPv6 Address] 2010:0:0:30::1
GigabitEthernet0/0/1               up              up
[IPv6 Address] 2001::1
GigabitEthernet0/0/2               up              up
[IPv6 Address] 2002::1
```

（2）R2 的配置。

```
[R2]display ipv6 interface brief
*down: administratively down
(l): loopback
(s): spoofing
Interface                          Physical        Protocol
GigabitEthernet0/0/0               up              up
[IPv6 Address] 2020::1
GigabitEthernet0/0/1               up              up
[IPv6 Address] 2001::2
GigabitEthernet0/0/2               up              up
[IPv6 Address] 2003::2
```

(3) R3 的配置。

```
[R3]display ipv6 interface brief
*down: administratively down
(l): loopback
(s): spoofing
Interface                        Physical              Protocol
GigabitEthernet0/0/0             up                    up
 [IPv6 Address] 2030::1
GigabitEthernet0/0/1             up                    up
 [IPv6 Address] 2003::3
GigabitEthernet0/0/2             up                    up
 [IPv6 Address] 2002::3
```

步骤三 查看路由表的信息

在路由器上使用 display ipv6 routing-table 命令，查看路由表信息。

(1) R1 的配置。

```
[R1]display ipv6 routing-table
Routing Table : Public
     Destinations : 14    Routes : 14

Destination  : ::1                        PrefixLength : 128
NextHop      : ::1                        Preference   : 0
Cost         : 0                          Protocol     : Direct
RelayNextHop : ::                         TunnelID     : 0x0
Interface    : InLoopBack0                Flags        : D

Destination  : 2001::                     PrefixLength : 64
NextHop      : 2001::1                    Preference   : 0
Cost         : 0                          Protocol     : Direct
RelayNextHop : ::                         TunnelID     : 0x0
Interface    : GigabitEthernet0/0/1       Flags        : D

Destination  : 2001::1                    PrefixLength : 128
NextHop      : ::1                        Preference   : 0
Cost         : 0                          Protocol     : Direct
RelayNextHop : ::                         TunnelID     : 0x0
Interface    : GigabitEthernet0/0/1       Flags        : D

Destination  : 2002::                     PrefixLength : 64
NextHop      : 2002::1                    Preference   : 0
Cost         : 0                          Protocol     : Direct
RelayNextHop : ::                         TunnelID     : 0x0
Interface    : GigabitEthernet0/0/2       Flags        : D
```

```
Destination   : 2002::1                    PrefixLength : 128
NextHop       : ::1                        Preference   : 0
Cost          : 0                          Protocol     : Direct
RelayNextHop  : ::                         TunnelID     : 0x0
Interface     : GigabitEthernet0/0/2       Flags        : D

Destination   : 2010:0:0:10::              PrefixLength : 64
NextHop       : 2010:0:0:10::1             Preference   : 0
Cost          : 0                          Protocol     : Direct
RelayNextHop  : ::                         TunnelID     : 0x0
Interface     : GigabitEthernet0/0/0.10    Flags        : D

Destination   : 2010:0:0:10::1             PrefixLength : 128
NextHop       : ::1                        Preference   : 0
Cost          : 0                          Protocol     : Direct
RelayNextHop  : ::                         TunnelID     : 0x0
Interface     : GigabitEthernet0/0/0.10    Flags        : D

Destination   : 2010:0:0:20::              PrefixLength : 64
NextHop       : 2010:0:0:20::1             Preference   : 0
Cost          : 0                          Protocol     : Direct
RelayNextHop  : ::                         TunnelID     : 0x0
Interface     : GigabitEthernet0/0/0.20    Flags        : D

Destination   : 2010:0:0:20::1             PrefixLength : 128
NextHop       : ::1                        Preference   : 0
Cost          : 0                          Protocol     : Direct
RelayNextHop  : ::                         TunnelID     : 0x0
Interface     : GigabitEthernet0/0/0.20    Flags        : D

Destination   : 2010:0:0:30::              PrefixLength : 64
NextHop       : 2010:0:0:30::1             Preference   : 0
Cost          : 0                          Protocol     : Direct
RelayNextHop  : ::                         TunnelID     : 0x0
Interface     : GigabitEthernet0/0/0.30    Flags        : D

Destination   : 2010:0:0:30::1             PrefixLength : 128
NextHop       : ::1                        Preference   : 0
Cost          : 0                          Protocol     : Direct
RelayNextHop  : ::                         TunnelID     : 0x0
Interface     : GigabitEthernet0/0/0.30    Flags        : D
```

```
Destination  : 2020::                    PrefixLength : 64
NextHop      : 2001::2                   Preference   : 60
Cost         : 0                         Protocol     : Static
RelayNextHop : ::                        TunnelID     : 0x0
Interface    : GigabitEthernet0/0/1      Flags        : RD

Destination  : 2030::                    PrefixLength : 64
NextHop      : 2002::3                   Preference   : 60
Cost         : 0                         Protocol     : Static
RelayNextHop : ::                        TunnelID     : 0x0
Interface    : GigabitEthernet0/0/2      Flags        : RD

Destination  : FE80::                    PrefixLength : 10
NextHop      : ::                        Preference   : 0
Cost         : 0                         Protocol     : Direct
RelayNextHop : ::                        TunnelID     : 0x0
Interface    : NULL0                     Flags        : D
```

（2）同理，可以查看 R2 和 R3 的路由表。

步骤四 测试计算机的连通性

通过 ping 命令，测试发现各计算机之间可以互相通信，如图 9-12、图 9-13 所示。

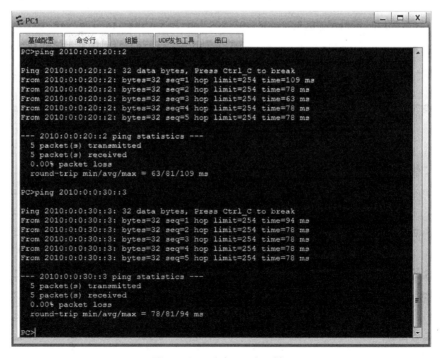

图 9-12 测试 PC 连通性 2

图 9-13　测试 PC 连通性 3

项目小结

本项目介绍了在网络内部署 IPv6 协议并实现 IPv6 网络互联互通的方法。通过本项目的学习，同学们掌握了接口 IPv6 地址的配置、计算机 IPv6 地址的配置以及基于 IPv6 路由的静态路由和单臂路由的配置方法。

项目 ⑩ WLAN 技术

项目目标

1. 掌握配置 AC 的无线网络的方法
2. 掌握配置 AC 的无线网络安全加密功能的方法

微课视频

WLAN 技术

任务 基于 WLAN 的公司无线网络配置

任务背景

以有线电缆或光纤作为传输介质的有线局域网应用广泛,但有线传输介质的铺设成本高,位置固定,移动性差。随着人们对网络的便携性和移动性的要求日益增强,传统的有线网络已经无法满足需求,WLAN 技术应运而生。目前,WLAN 已经成为一种经济、高效的网络接入方式。通过 WLAN 技术,用户可以方便地接入无线网络,并在无线网络覆盖区域内自由移动。

本项目将通过典型 AC+FIT AP 组网的模式,帮助学生理解构建基础 WLAN 网络的整体流程与配置。

某企业网络需要用户通过 WLAN 接入网络,以满足移动办公的最基本需求。网络拓扑图如图 10-1 所示,其中:

(1) AC 采用旁挂组网方式,AC 与 AP 处于同一个二层网络;

(2) AC 作为 DHCP 服务器给 AP 分配 IP 地址,S1 交换机作为 DHCP 服务器给接入的 STA 分配 IP 地址。

任务规划

配置思路如下:

(1) 配置有线网络侧互联互通。

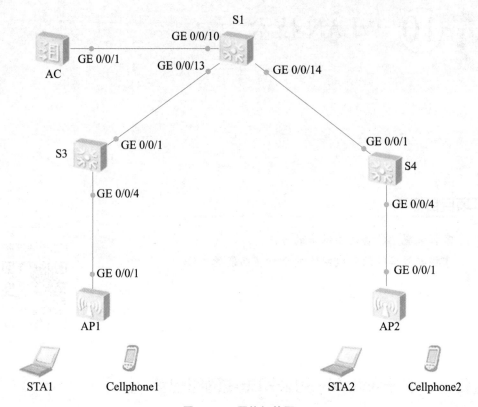

图 10-1 网络拓扑图

(2) 配置 AP 上线：
①创建 AP 组，用于将需要进行相同配置的 AP 都加入 AP 组，实现统一配置。
②配置 AC 的系统参数，包括国家码、AC 与 AP 之间通信的源接口。
③配置 AP 上线的认证方式并离线导入 AP，实现 AP 正常上线。
(3) 配置 WLAN 业务参数并下发给 AP，实现 STA 访问 WLAN 网络功能。
具体规划见表 10-1。

表 10-1 AC 数据规划表

配置项	配置参数
AP 管理 VLAN	VLAN 100
STA 业务 VLAN	VLAN 101
DHCP 服务器	AC 作为 DHCP 服务器为 AP 分配 IP 地址，S1 作为 DHCP 服务器为 STA 分配 IP 地址，STA 的默认网关为 192.168.101.254
AP 的 IP 地址池	192.168.100.1—192.168.100.253/24
STA 的 IP 地址池	192.168.101.1—192.168.101.253/24
AC 的源接口 IP 地址	VLANIF 100：192.168.100.254/24
AP 组	名称：ap-group1 引用模板：VAP 模板 huawei-wlan、域管理模板 default

续表

配置项	配置参数
域管理模板	名称：default
	国家码：中国（CN）
SSID 模板	名称：huawei-wlan
	SSID 名称：huawei-wlan
安全模板	名称：huawei-wlan
	安全策略：WPA－WPA2＋PSK＋AES
	密码：huawei-Datacom
VAP 模板	名称：huawei-wlan
	转发模式：直接转发
	业务 VLAN：VLAN 101
	引用模板：SSID 模板 huawei-wlan、安全模板 huawei-wlan

任务实施

步骤一 设备基本配置

在各交换机和 AC 上创建相应的 VLAN，将端口划分至相应的 VLAN。

（1）配置 S1 交换机接口。

```
<Huawei>system-view
[Huawei]sysname S1
[S1]vlan batch 100 101
Info: This operation may take a few seconds. Please wait for a moment...done.
[S1]interface GigabitEthernet 0/0/13
[S1-GigabitEthernet0/0/13]port link-type trunk
[S1-GigabitEthernet0/0/13]port trunk allow-pass vlan 100 101
[S1-GigabitEthernet0/0/13]quit
[S1]interface GigabitEthernet 0/0/14
[S1-GigabitEthernet0/0/14]port link-type trunk
[S1-GigabitEthernet0/0/14]port trunk allow-pass vlan 100 101
[S1-GigabitEthernet0/0/14]quit
[S1]interface GigabitEthernet 0/0/10
[S1-GigabitEthernet0/0/10]port link-type trunk
[S1-GigabitEthernet0/0/10]port trunk allow-pass vlan 100 101
[S1-GigabitEthernet0/0/10]quit
[S1]interface Vlanif 101
[S1-Vlanif101]ip address 192.168.101.254 24
```

（2）配置 AC 接口。

```
<AC6005>system-view
[AC6005]sysname AC
[AC]vlan batch 100 101
Info: This operation may take a few seconds. Please wait for a moment...done.
[AC]interface GigabitEthernet 0/0/1
[AC-GigabitEthernet0/0/10]port link-type trunk
[AC-GigabitEthernet0/0/10]port trunk allow-pass vlan 100 101
[AC-GigabitEthernet0/0/10]quit
[AC]interface Vlanif 100
[AC-Vlanif100]ip address 192.168.100.254 24
```

（3）配置 S3 交换机接口。

```
<Huawei>system-view
[Huawei]sysname S3
[S3]vlan batch 100 101
Info: This operation may take a few seconds. Please wait for a moment...done.
[S3]interface GigabitEthernet 0/0/1
[S3-GigabitEthernet0/0/1]port link-type trunk
[S3-GigabitEthernet0/0/1]port trunk allow-pass vlan 100 101
[S3-GigabitEthernet0/0/1]quit
[S3]interface GigabitEthernet 0/0/4
[S3-GigabitEthernet0/0/4]port link-type trunk
[S3-GigabitEthernet0/0/4]port trunk pvid vlan 100
[S3-GigabitEthernet0/0/4]port trunk allow-pass vlan 100 101
[S3-GigabitEthernet0/0/4]quit
```

（4）配置 S4 交换机接口。

```
<Huawei>system-view
[Huawei]sysname S4
[S4]vlan batch 100 101
Info: This operation may take a few seconds. Please wait for a moment...done.
[S4]interface GigabitEthernet0/0/1
[S4-GigabitEthernet0/0/1] port link-type trunk
[S4-GigabitEthernet0/0/1] port trunk allow-pass vlan 100 to 101
[S4-GigabitEthernet0/0/1]quit
[S4]interface GigabitEthernet0/0/4
[S4-GigabitEthernet0/0/4] port link-type trunk
[S4-GigabitEthernet0/0/4] port trunk pvid vlan 100
[S4-GigabitEthernet0/0/4] port trunk allow-pass vlan 100 to 101
[S4-GigabitEthernet0/0/4]quit
```

步骤二 DHCP 配置

在 S1 上创建 STA 接入时所使用的 IP 地址池，在 AC 上创建 AP 接入时所使用的 IP 地址池。S1 作为 STA 的 DHCP Server，AC 作为 AP 的 DHCP Server，分别配置 DHCP 服务。

（1）S1 上配置 DHCP。

```
[S1]dhcp enable
Info: The operation may take a few seconds. Please wait for a moment.done.
[S1]ip pool sta
Info:It's successful to create an IP address pool.
[S1-ip-pool-sta]network 192.168.101.0 mask 24
[S1-ip-pool-sta]gateway-list 192.168.101.254
[S1-ip-pool-sta]quit
[S1]interface Vlanif 101
[S1-Vlanif101]dhcp select global
[S1-Vlanif101]quit
```

（2）AC 上配置 DHCP。

```
[AC]dhcp enable
Info: The operation may take a few seconds. Please wait for a moment.done.
[AC]ip pool ap
Info: It is successful to create an IP address pool.
[AC-ip-pool-ap]network 192.168.100.254 mask 24
[AC-ip-pool-ap]gateway-list 192.168.100.254
[AC-ip-pool-ap]quit
[AC]interface Vlanif 100
[AC-Vlanif100]dhcp select global
[AC-Vlanif100]quit
```

步骤三 配置 AP 上线

（1）创建名为 ap-group1 的 AP 组，创建域管理模板，在域管理模板下配置 AC 的国家码，并在 AP 组下引用域管理模板。

```
[AC]wlan
[AC-wlan-view]ap-group name ap-group1
Info: This operation may take a few seconds. Please wait for a moment.done.
[AC-wlan-view]regulatory-domain-profile name default
[AC-wlan-regulate-domain-default]country-code cn
Info: The current country code is same with the input country code.
[AC-wlan-regulate-domain-default]quit
[AC-wlan-view]ap-group name ap-group1
[AC-wlan-ap-group-ap-group1]regulatory-domain-profile default
```

```
Warning: Modifying the country code will clear channel, power and antenna gain
configurations of the radio and reset the AP. Continue? [Y/N]:y
    [AC-wlan-ap-group-ap-group1]quit
```

域管理模板提供对 AP 的国家码、调优信道集合和调优带宽等的配置。缺省情况下，系统上存在名为 default 的域管理模板，因此当前进入了默认存在的 default 模板。

国家码用来标识 AP 射频所在的国家，不同国家码规定了不同的 AP 射频特性，包括 AP 的发送功率、支持的信道等。配置国家码是为了使 AP 的射频特性符合不同国家或区域的法律法规要求。缺省情况下，设备的国家码标识为"CN"。

（2）配置 AC 建立 CAPWAP 隧道的源接口，在 AC 上离线导入 AP，并将 AP 加入配置好的 AP 组"ap-group1"中。

①查看 AP1 和 AP2 的 mac 地址，用于引入，如图 10-2、图 10-3 所示。

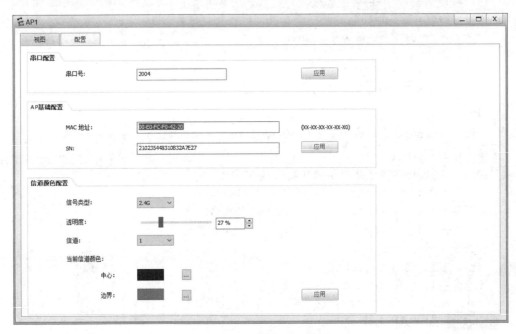

图 10-2　AP1 的 mac 地址

②引入 AP1 和 AP2 至 AC 组中。

```
[AC]capwap source interface Vlanif 100
[AC]wlan
[AC-wlan-view]ap auth-mode mac-auth 00E0-FCF0-4220
[AC-wlan-view]ap-id 0 ap-mac
[AC-wlan-ap-0]ap-name ap1
[AC-wlan-ap-0]ap-group ap-group1
Warning: This operation may cause AP reset. If the country code changes, it will
clear channel, power and antenna gain configurations of the radio, Whether to continue?
[Y/N]:y                    //需要输入 y 来确认继续
```

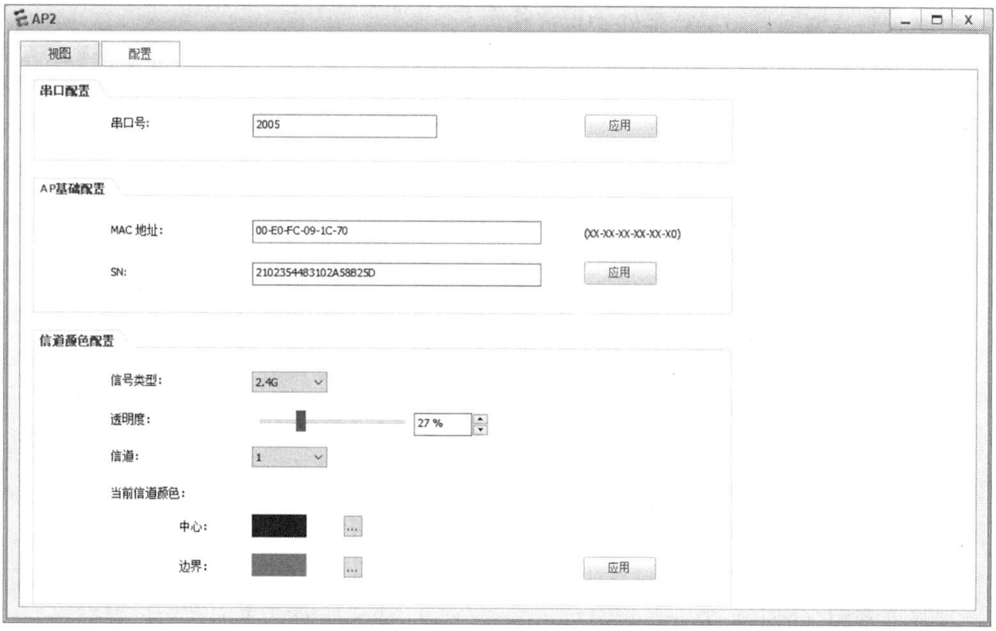

图 10-3　AP2 的 mac 地址

```
Info: This operation may take a few seconds. Please wait for a moment.. done.
[AC-wlan-ap-0]quit
[AC-wlan-view]ap-id 1 ap-mac 00E0-FC09-1C70
[AC-wlan-ap-1]ap-name ap2
[AC-wlan-ap-1]ap-group ap-group1
Warning: This operation may cause AP reset. If the country code changes, it will
clear channel, power and antenna gain configurations of the radio, Whether to continue?
[Y/N]:y              //需要输入 y 来确认继续
Info: This operation may take a few seconds. Please wait for a moment.. done.
[AC-wlan-ap-1]quit
```

（3）查看当前的 AP 信息。

```
[AC-wlan-view]display ap all
Info: This operation may take a few seconds. Please wait for a moment.done.
Total AP information:
nor  : normal           [2]
-----------------------------------------------------------------
ID   MAC           Name   Group      IP  Type  State  STA  Uptime
-----------------------------------------------------------------
0    00e0-fcf0-4220  ap1   ap-group1   -    -    idle   0    -
1    00e0-fc09-1c70  ap2   ap-group1   -    -    idle   0    -
-----------------------------------------------------------------
Total: 2
```

可以看到，此时两台 AP 都处于正常状态。

步骤四 配置 WLAN 业务参数

（1）创建名为"huawei-wlan"的安全模板，并配置安全策略。

```
[AC-wlan-view]security-profile name huawei-wlan
[AC-wlan-sec-prof-huawei-wlan]security wpa-wpa2 psk pass-phrase huawei-Datacom aes
[AC-wlan-sec-prof-huawei-wlan]quit
[AC]wlan
[AC-wlan-view]ssid-profile name huawei-wlan
[AC-wlan-ssid-prof-huawei-wlan]ssid huawei-wlan
Info: This operation may take a few seconds, please wait.done.
[AC-wlan-ssid-prof-huawei-wlan]quit
```

security psk 命令用来配置 WPA/WPA2 的预共享密钥认证和加密。

当前使用 WPA 和 WPA2 混合方式，用户终端使用 WPA 或 WPA2 都可以进行认证。预共享密钥（PSK）为 huawei-Datacom，通过 AES 加密算法加密用户数据。

（2）创建名为"huawei-wlan"的 VAP 模板，配置业务数据转发模式、业务 VLAN，并且引用安全模板和 SSID 模板。

```
[AC]wlan
[AC-wlan-view]vap-profile name huawei-wlan
[AC-wlan-vap-prof-huawei-wlan]forward-mode direct-forward
[AC-wlan-vap-prof-huawei-wlan]service-vlan vlan-id 101
Info: This operation may take a few seconds, please wait.done.
[AC-wlan-vap-prof-huawei-wlan]security-profile huawei-wlan
Info: This operation may take a few seconds, please wait.done.
[AC-wlan-vap-prof-huawei-wlan]ssid-profile huawei-wlan
Info: This operation may take a few seconds, please wait.done.
[AC-wlan-vap-prof-huawei-wlan]quit
```

（3）配置 AP 组引用 VAP 模板，AP 上射频 0 和射频 1 都使用 VAP 模板"huawei-wlan"的配置。

```
[AC]wlan
[AC-wlan-view]ap-group name ap-group1
[AC-wlan-ap-group-ap-group1]vap-profile huawei-wlan wlan 1 radio all
Info: This operation may take a few seconds, please wait...done.
[AC-wlan-ap-group-ap-group1]quit
```

稍等片刻后，AP1 和 AP2 启动成功，如图 10-4 所示：

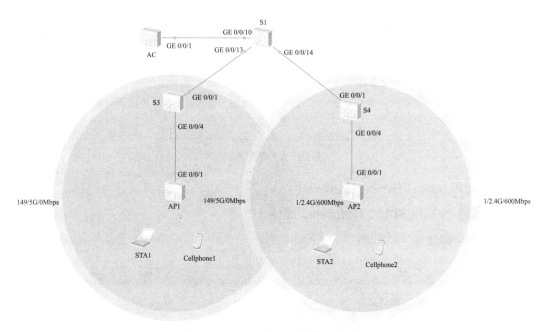

图 10-4　AP1 和 AP2 发射无线信号

任务验证

步骤一　分别用 STA1、Cellphone1 连接 AP1

STA1 连接 AP1 如图 10-5、图 10-6 所示。

图 10-5　STA1 连接无线

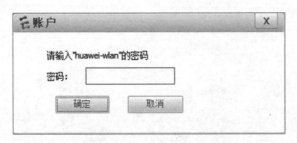

图 10-6　需要密码

输入密码 huawei-Datacom 连接成功。

步骤二　**分别用 STA2、Cellphone2 连接 AP2**

Cellphone2 连接无线如图 10-7 所示。

图 10-7　Cellphone2 连接无线

输入密码 huawei-Datacom 后连接成功。此时，移动设备均已连接无线，如图 10-8 所示。

步骤三　**查看设备的 IP 地址**

查看 STA1 的 IP 地址如图 10-9 所示。

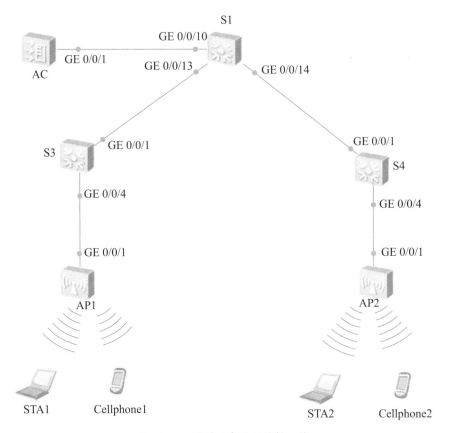

图 10-8 移动设备均已连接无线

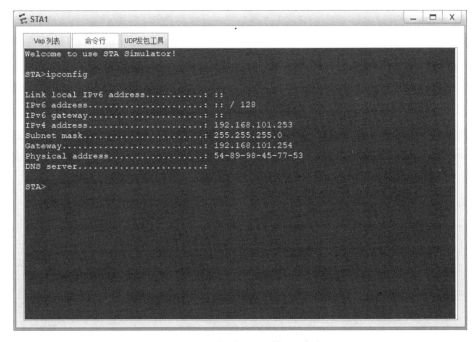

图 10-9 查看 STA1 的 IP 地址

可以看到，STA1 通过 DHCP 功能自动分配到了地址池中的一个 IP 地址。

项目小结

无线办公需求日益增加，本项目讲解了如何在公司已有的有线网络之上部署无线网络，使各部门用户可以接入无线网，实现移动办公。通过本项目的学习，同学们可以掌握 WLAN 中设备 AC 及 AP 的基本配置命令，能够完成小型企业无线网络的独自部署，拥有管理企业 WLAN 的基础能力。

项目 11 网络自动化运维

项目目标

1. 掌握编写 python 修改密码脚本的方法
2. 掌握编写 python 备份脚本的方法
3. 掌握配置计划任务的方法

微课视频

网络自动化运维

任务　某公司网络自动化运维配置

任务背景

某公司有三大办公区域，各区域之间使用路由器互联。技术部、财务部、市场部的路由器分别为 R1、R2、R3，路由器需配置单区域 OSPF 动态路由，使所有计算机能够互相访问。公司因安全考虑所需，需要统一修改所有网络设备管理密码，并进行每天的自动备份。网络拓扑图如图 11-1 所示，具体要求如下：

（1）各路由器之间配置 OSPF 路由实现网络互联互通；

（2）各路由器上启用 SNMP 协议，实现路由器可网管，对所有设备统一修改密码并进行每天的自动备份；

（3）测试计算机和路由器的 IP 地址及接口信息如图 11-1 所示。

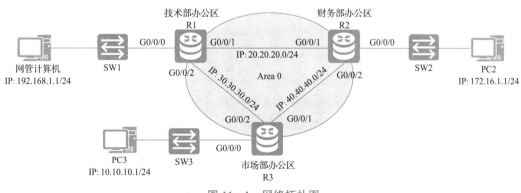

图 11-1　网络拓扑图

路由交换技术

📋 任务规划

三大办公区域能相互通信，路由器配置单区域 OSPF 动态路由，使所有计算机均能互访。所有网络设备开启 ssh，技术部设有网管计算机，确保网管计算机与网络设备正常通信。

配置步骤如下：
（1）配置路由器接口；
（2）部署单区域 OSPF 网络；
（3）路由器配置 ssh 登录；
（4）网管服务器安装模块；
（5）编写 python 修改密码脚本；
（6）编写 python 备份脚本；
（7）配置计划任务；
（8）配置各计算机的 IP 地址。

具体规划见表 11-1。

表 11-1 接口规划表

本端设备	接口	端口 IP 地址	对端设备
R1	G0/0/0	192.168.1.2/24	PC1
R1	G0/0/1	20.20.20.1/24	R2
R1	G0/0/2	30.30.30.1/24	R3
R2	G0/0/0	172.16.1.2/24	PC2
R2	G0/0/1	20.20.20.2/24	R1
R2	G0/0/2	40.40.40.2/24	R3
R3	G0/0/0	10.10.10.2/24	PC3
R3	G0/0/1	40.40.40.1/24	R2
R3	G0/0/2	30.30.30.10/24	R1

🔧 任务实施

步骤一 配置路由器接口

（1）R1 的配置。

```
[Huawei]system-view
[Huawei]sysname R1
[R1]interface GigabitEthernet 0/0/0
[R1-GigabitEthernet0/0/0]ip address 192.168.1.2 255.255.255.0
[R1]interface GigabitEthernet 0/0/1
[R1-GigabitEthernet0/0/1]ip address 20.20.20.1 255.255.255.0
[R1]interface GigabitEthernet 0/0/2
[R1-GigabitEthernet0/0/2]ip address 30.30.30.1 255.255.255.0
```

(2) R2 的配置。

```
[Huawei]system-view
[Huawei]sysname R2
[R2]interface GigabitEthernet 0/0/0
[R2-GigabitEthernet0/0/0]ip address 172.16.1.2 255.255.255.0
[R2]interface GigabitEthernet 0/0/1
[R2-GigabitEthernet0/0/1]ip address 20.20.20.2 255.255.255.0
[R2]interface GigabitEthernet 0/0/2
[R2-GigabitEthernet0/0/2]ip address 40.40.40.2 255.255.255.0
```

(3) R3 的配置。

```
[Huawei]system-view
[Huawei]sysname R3
[R3]interface GigabitEthernet 0/0/0
[R3-GigabitEthernet0/0/0]ip address 10.10.10.2 255.255.255.0
[R3]interface GigabitEthernet 0/0/1
[R3-GigabitEthernet0/0/1]ip address 40.40.40.1 255.255.255.0
[R3]interface GigabitEthernet 0/0/2
[R3-GigabitEthernet0/0/2]ip address 30.30.30.2 255.255.255.0
```

步骤二 部署单区域 OSPF 网络

首先创建并运行 OSPF，然后创建区域并进入 OPSF 区域视图，最后指定运行 OSPF 协议的接口和接口所属的区域。

(1) R1 的配置。

```
[R1]ospf 1
[R1-ospf-1]area 0
[R1-ospf-1-area-0.0.0.0]network 192.168.1.0 0.0.0.255
[R1-ospf-1-area-0.0.0.0]network 20.20.20.0 0.0.0.255
[R1-ospf-1-area-0.0.0.0]network 30.30.30.0 0.0.0.255
```

(2) R2 的配置。

```
[R2]ospf 1
[R2-ospf-1]area 0
[R2-ospf-1-area-0.0.0.0]network 172.16.1.0 0.0.0.255
[R2-ospf-1-area-0.0.0.0]network 20.20.20.0 0.0.0.255
[R2-ospf-1-area-0.0.0.0]network 40.40.40.0 0.0.0.255
```

(3) R3 的配置。

```
[R3]ospf 1
[R3-ospf-1]area 0
[R3-ospf-1-area-0.0.0.0]network 10.10.10.0 0.0.0.255
[R3-ospf-1-area-0.0.0.0]network 40.40.40.0 0.0.0.255
[R3-ospf-1-area-0.0.0.0]network 30.30.30.0 0.0.0.255
```

步骤三 路由器配置 ssh 登录

(1) R1 的配置。

```
[R1]rsa local-key-pair create
The key name will be: Host
%RSA keys defined for Host already exist.
Confirm to replace them? (y/n)[n]:y
The range of public key size is (512 ~2048).
NOTES: If the key modulus is greater than 512,
       It will take a few minutes.
Input the bits in the modulus[default =512]:2048
[R1]aaa
[R1-aaa]local-user admin password cipher 123456
[R1-aaa]local-user admin privilege level 3
[R1-aaa]local-user admin service-type ssh
[R1-aaa]stelnet server enable
Info: Succeeded in starting the STELNET server.
[R1]ssh user admin authentication-type password
 Authentication type setted, and will be in effect next time
[R1]user-interface vty 0 4
[R1-ui-vty0-4]authentication-mode aaa
[R1-ui-vty0-4]protocol inbound ssh
```

(2) R2 的配置。

```
[R2]rsa local-key-pair create
The key name will be: Host
%RSA keys defined for Host already exist.
Confirm to replace them? (y/n)[n]:y
The range of public key size is (512 ~2048).
NOTES: If the key modulus is greater than 512,
       It will take a few minutes.
Input the bits in the modulus[default =512]:2048
[R2]aaa
[R2-aaa]local-user admin password cipher 123456
[R2-aaa]local-user admin privilege level 3
[R2-aaa]local-user admin service-type ssh
[R2-aaa]stelnet server enable
Info: Succeeded in starting the STELNET server.
[R2]ssh user admin authentication-type password
 Authentication type setted, and will be in effect next time
[R2]user-interface vty 0 4
[R2-ui-vty0-4]authentication-mode aaa
[R2-ui-vty0-4]protocol inbound ssh
```

（3）R3 的配置。

```
[R3]rsa local-key-pair create
The key name will be: Host
% RSA keys defined for Host already exist.
Confirm to replace them? (y/n) [n]:y
The range of public key size is (512~2048).
NOTES: If the key modulus is greater than 512,
       It will take a few minutes.
Input the bits in the modulus[default=512]:2048
[R3]aaa
[R3-aaa]local-user admin password cipher 123456
[R3-aaa]local-user admin privilege level 3
[R3-aaa]local-user admin service-type ssh
[R3-aaa]stelnet server enable
Info: Succeeded in starting the STELNET server.
[R3]ssh user admin authentication-type password
 Authentication type setted, and will be in effect next time
[R3]user-interface vty 0 4
[R3-ui-vty0-4]authentication-mode aaa
[R3-ui-vty0-4]protocol inbound ssh
```

步骤四 网管服务器安装模块

网管计算机联网状态下安装模块 paramiko。

```
[root@manage ~]# curl "https://bootstrap.pypa.io/get-pip.py" -o "get-pip.py"
[root@manage ~]#python get-pip.py
[root@manage ~]#pip install paramiko
```

步骤五 编写 python 修改密码脚本

编写 python 脚本 "changepassword.py"，实现对 R1、R2、R3 的密码修改。

```
[root@manage ~]#vi changepassword.py
##导入 paramiko、time、getpass 模块
#!/usr/bin/python
import paramiko
import time
import getpass
##通过 raw_input()函数获取用户输入的 SSH 用户名并赋值给 username
username=raw_input('Username:')
```

```
##通过 getpass 模块中的 getpass()函数获取用户输入字符串作为密码赋值给 password
password = getpass.getpass('Password:')
for i in ["192.168.1.2","172.16.1.2","10.10.10.2"]:
    ip=str(i)
    ssh_client=paramiko.SSHClient()
    ssh_client.set_missing_host_key_policy(paramiko.AutoAddPolicy())
    ssh_client.connect(hostname=ip,username=username,password=password)
    command=ssh_client.invoke_shell()
##调度交换机命令行执行命令
    command.send("system-view" +"\n")
    command.send("user-interface console 0"+"\n")
    command.send("set authentication password cipher 234567"+"\n")
##更改登录密码结束后,返回用户视图并保存配置
    command.send("return"+"\n")
    command.send("save"+"\n")
    command.send("Y"+"\n")
    command.send("\n")
##暂停 2 秒,并将命令执行过程赋值给 output 对象,通过 print output 语句回显内容
    time.sleep(2)
    output=command.recv(65535)
    print output
##退出 SSH
ssh_client.close()
```

步骤六 编写 python 备份脚本

在网管计算机创建备份交换机运行配置的脚本"backup.py"。

```
[root@manage ~]#vi backup.py
##导入 paramiko、time、datetime 等模块
#!/usr/bin/python
import paramiko
import time
from datetime import datetime
##设置 SSH 用户名和密码
username ="admin"
password ="234567"
##通过 for 语句遍历 i 的值为 1/2/3/4,结合 ip="192.168.100." +str(i)语句循环 SSH 登录交换机设备
for i in range(1,5):
    ip="192.168.100." +str(i)
    ssh_client=paramiko.SSHClient()
```

```
    ssh_client.set_missing_host_key_policy(paramiko.AutoAddPolicy())
    ssh_client.connect(hostname=ip,username=username,password=password)
    command=ssh_client.invoke_shell()
##提示SSH登录成功
    print "ssh "+ip+" successfully"
##设置回显内容不分屏显示
    command.send("screen-length 0 temporary "+"\n")
##获取交换机运行配置
    output=(command.send("display current-configuration" +"\n")
##程序暂停2秒
    time.sleep(2)
##读取当前时间
    now=datetime.now()
##打开备份文件

backup=open("/root/backup/"+str(now.year)+"-"+str(now.month)+"-"+str(now.day)+"-"+ip+".txt","a+")
##提示正在备份
    print "backuping"
##将查询运行配置的回显内容赋值给recv这个对象
    recv=command.recv(65535)
##将回显内容写入backup这个对象,相当于写入了备份文件中
    backup.write(recv)
##关闭打开的文件
    backup.close()
##结束,断开SSH连接
ssh_client.close()
```

步骤七 配置计划任务

配置计划任务实现每天凌晨1点自动执行脚本进行备份。

```
[root@manage ~]#vi /etc/crontab
##在文件末尾填入下列内容后退出
00 1 * * * root python /root/backup.py
[root@manage ~]#mkdir /root/backup
[root@manage ~]#systemctl restart crond
[root@manage ~]#systemctl enable crond
```

步骤八 配置各计算机的IP地址

各计算机的IP地址配置如图11-2、图11-3所示。

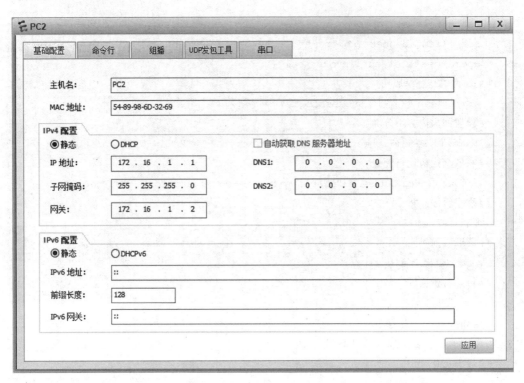

图 11-2　PC2 IP 地址配置图

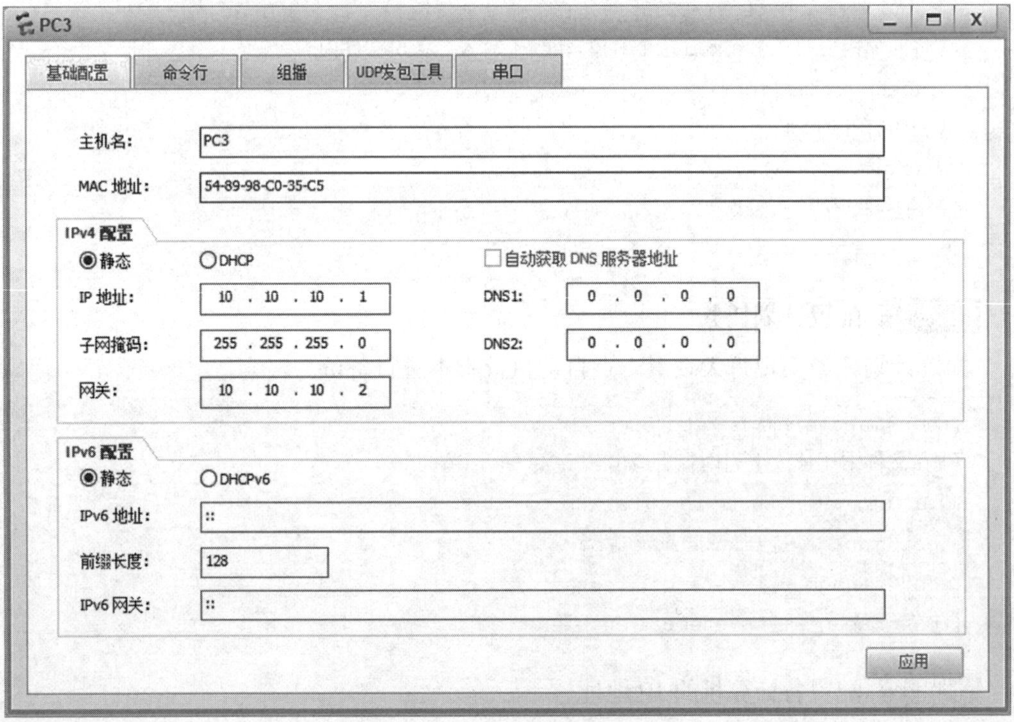

图 11-3　PC3 IP 地址配置图

任务验证

步骤一 执行 changepassword.py

执行 changepassword.py，查看回显内容。

```
[root@manage ~]#./changepassword.py
Username:admin
Password:

-----------------------------------------------------------------
  User last login information:
-----------------------------------------------------------------
  Access Type : SSH
  IP-Address  : 192.168.1.130 ssh
  Time        : 2020-02-29 10:31:35-08:00
-----------------------------------------------------------------
<R1>system-view
Enter system view, return user view with Ctrl+Z.
[R1]user-interface console 0
[R1-ui-console0]set authentication password cipher 234567
[R1-ui-console0]return
<R1>save
  The current configuration will be written to the device.
  Are you sure to continue? (y/n)[n]:Y
  It will take several minutes to save configuration file, please wait...


-----------------------------------------------------------------
  User last login information:
-----------------------------------------------------------------
  Access Type : SSH
  IP-Address  : 192.168.1.130 ssh
  Time        : 2020-02-29 10:31:38-08:00
-----------------------------------------------------------------
<R2>system-view
Enter system view, return user view with Ctrl+Z.
[R2]user-interface console 0
[R2-ui-console0]set authentication password cipher 234567
[R2-ui-console0]return
<R2>save
  The current configuration will be written to the device.
  Are you sure to continue? (y/n)[n]:Y
  It will take several minutes to save configuration file, please wait...
```

```
  ----------------------------------------------------------------
    User last login information:
  ----------------------------------------------------------------
    Access Type : SSH
    IP-Address  : 192.168.1.130 ssh
    Time        : 2020-02-29 10:31:41-08:00
  ----------------------------------------------------------------
<R3>system-view
Enter system view, return user view with Ctrl+Z.
[R3]user-interface console 0
[R3-ui-console0]set authentication password cipher 234567
[R3-ui-console0]return
<R3>save
  The current configuration will be written to the device.
  Are you sure to continue? (y/n)[n]:Y
  It will take several minutes to save configuration file, please wait...
```

步骤二 计划任务执行后查看备份文件

查看/root/backup 目录下的文件。

```
[root@manage ~]#cd /root/backup
[root@manage backup]#ls
2020-2-28-10.10.10.2.txt   2020-2-28-172.16.1.2.txt   2020-2-28-192.168.1.2.txt
[root@manage backup]#ll
total 12
-rw-r--r--. 1 root root 1786 Feb 28 1:00 2020-2-28-10.10.10.2.txt
-rw-r--r--. 1 root root 1809 Feb 28 1:00 2020-2-28-172.16.1.2.txt
-rw-r--r--. 1 root root 1762 Feb 28 1:00 2020-2-28-192.168.1.2.txt
```

查看详细内容。

```
[root@manage backup]#cat 2020-2-28-10.10.10.2.txt


  ----------------------------------------------------------------
    User last login information:
  ----------------------------------------------------------------
    Access Type: SSH
    IP-Address : 192.168.1.130 ssh
    Time       : 2020-02-29 10:32:24-08:00
  ----------------------------------------------------------------
<R3>screen-length 0 temporary
Info: The configuration takes effect on the current user terminal interface only.
```

```
<R3>display current-configuration
[V200R003C00]
#
 sysname R3
#
 snmp-agent local-engineid 800007DB03000000000000
 snmp-agent
#
 clock timezone China-Standard-Time minus 08:00:00
#
 portal local-server load flash:/portalpage.zip
#
 drop illegal-mac alarm
#
 wlan ac-global carrier id other ac id 0
#
 set cpu-usage threshold 80 restore 75
#
aaa
 authentication-scheme default
 authorization-scheme default
 accounting-scheme default
 domain default
 domain default_admin
 local-user admin password cipher %$%$YgN! G* Q* }0tjsqA"g~X(T{]!%$%$
 local-user admin privilege level 3
 local-user admin service-type ssh
#
firewall zone Local
 priority 15
#
interface GigabitEthernet0/0/0
 ip address 10.10.10.2 255.255.255.0
#
interface GigabitEthernet0/0/1
 ip address 40.40.40.2 255.255.255.0
#
interface GigabitEthernet0/0/2
 ip address 30.30.30.2 255.255.255.0
```

```
#
interface NULL0
#
ospf 1
  area 0.0.0.0
    network 10.10.10.0 0.0.0.255
    network 30.30.30.0 0.0.0.255
    network 40.40.40.0 0.0.0.255
#
  stelnet server enable
#
user-interface con 0
  authentication-mode password
  set authentication password cipher %$%$v:.{Vo~Vt;s;grBK&HD9,%S$wJDM)zzk69v.\&X&+%jX%S',%$%$
  user-interface vty 0 4
    authentication-mode aaa[root@manage backup]#
```

可以观察到，路由器整个配置信息已备份下来。

项目小结

企业网络结构复杂、设备众多，运维成本较大。本项目讲解了如何利用 python 实现网络的自动化运维，通过在一台主管设备上引入 python 脚本，便可管理网络中所有设备，免除了网络管理员进行大量重复工作带来的时间和精力消耗，节约人力、物力，大大降低了企业网络的运维成本。

参考文献

[1] 孙良旭，李林林，吴建胜．路由交换技术［M］．2版．北京：清华大学出版社，2016．

[2] 姜丹丹，等．路由与交换技术实战入门与提高［M］．北京：科学出版社，2012．

[3] Allan Reid，Jim Lorenz，Cheryl Schmidt. CCNA Discovery：企业中的路由和交换简介［M］．思科系统公司，译．北京：人民邮电出版社，2009．

[4] 斯桃枝．路由与交换技术［M］．北京：北京大学出版社，2008．

[5] 史律，钱亮，陈永．交换与路由技术［M］．北京：高等教育出版社，2022．

[6] 张保通，李伟红．网络互连技术——路由、交换与远程访问［M］．2版．北京：中国水利水电出版，2008．

[7] 韩立刚，李圣春，韩利辉．华为HCNA路由与交换学习指南［M］．北京：人民邮电出版社，2019．

[8] 朱仕耿．HCNP路由交换学习指南［M］．北京：人民邮电出版社，2017．

[9] 华为技术有限公司．网络系统建设与运维（中级）［M］．北京：人民邮电出版社，2020．

[10] 刘丹宁，田果，韩士良．路由与交换技术［M］．北京：人民邮电出版社，2023．

附 录

附录 A 常用命令

PC 命令：

PCA login: root	;使用 root 用户
password: linux	;口令是 linux
# shutdown -h now	;关机
# init 0	;关机
# logout	;用户注销
# login	;用户登录
# ifconfig	;显示 IP 地址
# ifconfig eth0 \<ip address\>netmask \<netmask\>	;设置 IP 地址
# ifconfig eht0 \<ip address\>netmask \<netmask\>down	;禁用 IP 地址
# route add 0.0.0.0 gw \<ip\>	;设置网关
# route del 0.0.0.0 gw \<ip\>	;删除网关
# route add default gw \<ip\>	;设置网关
# route del default gw \<ip\>	;删除网关
# route	;显示网关
# ping \<ip\>	;发 ECHO 包
# telnet \<ip\>	;远程登录

交换机命令：

[Quidway]dis cur	;显示当前配置
[Quidway]display current-configuration	;显示当前配置
[Quidway]display interfaces	;显示接口信息
[Quidway]display vlan all	;显示路由信息
[Quidway]display version	;显示版本信息
[Quidway]super password	;修改特权用户密码
[Quidway]sysname	;交换机命名
[Quidway]interface ethernet 0/1	;进入接口视图
[Quidway]interface vlan x	;进入接口视图

```
[Quidway-Vlan-interfacex]ip address 10.65.1.1 255.255.0.0    ;配置 VLAN 的 IP 地址
[Quidway]ip route-static 0.0.0.0 0.0.0.0 10.65.1.2           ;静态路由=网关
[Quidway]rip                                                 ;三层交换支持
[Quidway]local-user ftp
[Quidway]user-interface vty 0 4                              ;进入虚拟终端
[S3026-ui-vty0-4]authentication-mode password                ;设置口令模式
[S3026-ui-vty0-4]set authentication-mode password simple 222
                                                             ;设置口令
[S3026-ui-vty0-4]user privilege level 3                      ;用户级别
[Quidway]interface ethernet 0/1                              ;进入端口模式
[Quidway]int e0/1                                            ;进入端口模式
[Quidway-Ethernet0/1]duplex {half|full|auto}                 ;配置端口工作状态
[Quidway-Ethernet0/1]speed {10|100|auto}                     ;配置端口工作速率
[Quidway-Ethernet0/1]flow-control                            ;配置端口流控
[Quidway-Ethernet0/1]mdi {across|auto|normal}                ;配置端口平接扭接
[Quidway-Ethernet0/1]port link-type {trunk|access|hybrid}
                                                             ;设置端口工作模式
[Quidway-Ethernet0/1]port access vlan 3                      ;当前端口加入 VLAN
[Quidway-Ethernet0/2]port trunk permit vlan {ID|All}         ;设 trunk 允许的 VLAN
[Quidway-Ethernet0/3]port trunk pvid vlan 3                  ;设置 trunk 端口的 PVID
[Quidway-Ethernet0/1]undo shutdown                           ;激活端口
[Quidway-Ethernet0/1]shutdown                                ;关闭端口
[Quidway-Ethernet0/1]quit                                    ;返回
[Quidway]vlan 3                                              ;创建 VLAN
[Quidway-vlan3]port ethernet 0/1                             ;在 VLAN 中增加端口
[Quidway-vlan3]port e0/1                                     ;简写方式
[Quidway-vlan3]port ethernet 0/1 to ethernet 0/4             ;在 VLAN 中增加端口
[Quidway-vlan3]port e0/1 to e0/4                             ;简写方式
[Quidway]monitor-port <interface_type interface_num>         ;指定镜像端口
[Quidway]port mirror <interface_type interface_num>          ;指定被镜像端口
[Quidway]port mirror int_list observing-port int_type int_num
                                                             ;指定镜像和被镜像
[Quidway]description string                                  ;指定 VLAN 描述字符
[Quidway]description                                         ;删除 VLAN 描述字符
[Quidway]display vlan [vlan_id]                              ;查看 VLAN 设置
[Quidway]stp {enable|disable}                                ;设置生成树,默认关闭
[Quidway]stp priority 4096                                   ;设置交换机的优先级
[Quidway]stp root {primary|secondary}                        ;设置为根或根的备份
[Quidway-Ethernet0/1]stp cost 200                            ;设置交换机端口的花费
[Quidway]link-aggregation e0/1 to e0/4 ingress|both          ;端口的聚合
```

```
[Quidway]undo link-aggregation e0/1|all              ;始端口为通道号
[SwitchA-vlanx]isolate-user-vlan enable              ;设置主 vlan
[SwitchA]isolate-user-vlan <x>secondary <list>       ;设置主 vlan 包括的子 vlan
[Quidway-Ethernet0/2]port hybrid pvid vlan <id>      ;设置 vlan 的 pvid
[Quidway-Ethernet0/2]port hybrid pvid               ;删除 vlan 的 pvid
[Quidway-Ethernet0/2]port hybrid vlan vlan_id_list untagged
                                                    ;设置无标识的 vlan
```

如果包的 vlan id 与 pvid 一致，则去掉 vlan 信息. 默认 PVID=1。
所以设置 pvid 为所属 vlan id, 设置可以互通的 vlan 为 untagged。

路由器命令：

```
[Quidway]display version                             ;显示版本信息
[Quidway]display current-configuration               ;显示当前配置
[Quidway]display interfaces                          ;显示接口信息
[Quidway]display ip route                            ;显示路由信息
[Quidway]sysname aabbcc                              ;更改主机名
[Quidway]super passwrod 123456                       ;设置口令
[Quidway]interface serial0                           ;进入接口
[Quidway-serial0]ip address <ip><mask|mask_len>      ;配置端口 IP 地址
[Quidway-serial0]undo shutdown                       ;激活端口
[Quidway]link-protocol hdlc                          ;绑定 hdlc 协议
[Quidway]user-interface vty 0 4
[Quidway-ui-vty0-4]authentication-mode password
[Quidway-ui-vty0-4]set authentication-mode password simple 222
[Quidway-ui-vty0-4]user privilege level 3
[Quidway-ui-vty0-4]quit
[Quidway]debugging hdlc all serial0                  ;显示所有信息
[Quidway]debugging hdlc event serial0                ;调试事件信息
[Quidway]debugging hdlc packet serial0               ;显示包的信息
```

静态路由：

```
[Quidway]ip route-static        <ip><mask>{interface number|nexthop}[value][reject|blackhole]
例如：
[Quidway]ip route-static 129.1.0.0 16 10.0.0.2
[Quidway]ip route-static 129.1.0.0 255.255.0.0 10.0.0.2
[Quidway]ip route-static 129.1.0.0 16 Serial 2
[Quidway]ip route-static 0.0.0.0 0.0.0.0  10.0.0.2
```

动态路由:

[Quidway]rip   ;设置动态路由
[Quidway]rip work   ;设置工作允许
[Quidway]rip input   ;设置入口允许
[Quidway]rip output   ;设置出口允许
[Quidway-rip]network 1.0.0.0   ;设置交换路由网络
[Quidway-rip]network all   ;设置与所有网络交换
[Quidway-rip]peer ip-address   ;
[Quidway-rip]summary   ;路由聚合
[Quidway]rip version 1   ;设置工作在版本 1
[Quidway]rip version 2 multicast   ;设版本 2,多播方式
[Quidway-Ethernet0]rip split-horizon   ;水平分隔
[Quidway]router id A.B.C.D   ;配置路由器的 ID
[Quidway]ospf enable   ;启动 OSPF 协议
[Quidway-ospf]import-route direct   ;引入直联路由
[Quidway-Serial0]ospf enable area <area_id>   ;配置 OSPF 区域

标准访问列表命令格式:

acl <acl-number> [match-order config|auto]   ;默认前者顺序匹配
rule [normal|special]{permit|deny} [source source-addr source-wildcard|any]
例如:
[Quidway]acl 10
[Quidway-acl-10]rule normal permit source 10.0.0.0 0.0.0.255
[Quidway-acl-10]rule normal deny source any

扩展访问控制列表配置命令:

配置 TCP/UDP 协议的扩展访问列表:
rule {normal|special}{permit|deny}{tcp|udp}source {<ip wild>|any}destination <ip wild>|any}

[operate]

配置 ICMP 协议的扩展访问列表:
rule {normal|special}{permit|deny}icmp source {<ip wild>|any}destination {<ip wild>|any}

[icmp-code] [logging]

扩展访问控制列表操作符的含义:

equal portnumber   ;等于
greater-than portnumber   ;大于
less-than portnumber   ;小于
not-equal portnumber   ;不等
range portnumber1 portnumber2   ;区间

扩展访问控制列表举例：

[Quidway]acl 101

[Quidway-acl-101]rule deny souce any destination any

[Quidway-acl-101]rule permit icmp source any destination any icmp-type echo

[Quidway-acl-101]rule permit icmp source any destination any icmp-type echo-reply

[Quidway]acl 102

[Quidway-acl-102]rule permit ip source 10.0.0.1 0.0.0.0 destination 202.0.0.1 0.0.0.0

[Quidway-acl-102]rule deny ip source any destination any

[Quidway]acl 103

[Quidway-acl-103]rule permit tcp source any destination 10.0.0.1 0.0.0.0 destination-port equal ftp

[Quidway-acl-103]rule permit tcp source any destination 10.0.0.2 0.0.0.0 destination-port equal www

[Quidway]firewall enable

[Quidway]firewall default permit|deny

[Quidway]int e0

[Quidway-Ethernet0]firewall packet-filter 101 inbound|outbound

地址转换配置举例：

[Quidway]firewall enable

[Quidway]firewall default permit

[Quidway]acl 101   ;内部指定主机可以进入 e0

[Quidway-acl-101]rule deny ip source any destination any

[Quidway-acl-101]rule permit ip source 129.38.1.1 0 destination any

[Quidway-acl-101]rule permit ip source 129.38.1.2 0 destination any

[Quidway-acl-101]rule permit ip source 129.38.1.3 0 destination any

[Quidway-acl-101]rule permit ip source 129.38.1.4 0 destination any

[Quidway-acl-101]quit

[Quidway]int e0

[Quidway-Ethernet0]firewall packet-filter 101 inbound

[Quidway]acl 102   ;外部特定主机和大于 1024
端口的数据包允许进入 S0

[Quidway-acl-102]rule deny ip source any destination any

[Quidway-acl-102]rule permit tcp source 202.39.2.3 0 destination 202.38.160.1 0

[Quidway-acl-102]rule permit tcp source any destination 202.38.160.1 0 destination-port great-than 1024

[Quidway-acl-102]quit

[Quidway]int s0

[Quidway-Serial0]firewall packet-filter 102 inbound   ;设置防火墙策略
[Quidway-Serial0]nat outbound 101 interface   ;是 Easy ip,将 acl 101 允许的 IP 从本接口出时变换源地址

内部服务器地址转换配置命令（静态 nat）：

nat server global <ip>[port] inside <ip>port [protocol]   ;global_port 不写时使用 inside_port
[Quidway-Serial0]nat server global 202.38.160.1 inside 129.38.1.1 ftp tcp
[Quidway-Serial0]nat server global 202.38.160.1 inside 129.38.1.2 telnet tcp
[Quidway-Serial0]nat server global 202.38.160.1 inside 129.38.1.3 www tcp
设有公网 IP:202.38.160.101~202.38.160.103 可以使用。   ;对外访问
[Quidway]nat address-group 202.38.160.101 202.38.160.103 pool1   ;建立地址池
[Quidway]acl 1
[Quidway-acl-1]rule permit source 10.110.10.0 0.0.0.255   ;指定允许的内部网络
[Quidway-acl-1]rule deny source any
[Quidway-acl-1]int serial 0
[Quidway-Serial0]nat outbound 1 address-group pool1   ;在 s0 口从地址池取出 IP 对外访问
[Quidway-Serial0]nat server global 202.38.160.101 inside 10.110.10.1 ftp tcp
[Quidway-Serial0]nat server global 202.38.160.102 inside 10.110.10.2 www tcp
[Quidway-Serial0]nat server global 202.38.160.102 8080 inside 10.110.10.3 www tcp
[Quidway-Serial0]nat server global 202.38.160.103 inside 10.110.10.4 smtp udp

PPP 设置：

[Quidway-s0]link-protocol ppp   ;默认的协议
PPP 验证：
主验方:pap|chap
[Quidway]local-user q2 password {simple|cipher} hello   ;路由器 1
[Quidway]interface serial 0
[Quidway-serial0]ppp authentication-mode {pap|chap}
[Quidway-serial0]ppp chap user q1   ;pap 时,没有此句
pap 被验方：
[Quidway]interface serial 0   ;路由器 2
[Quidway-serial0]ppp pap local-user q2 password {simple|cipher} hello
chap 被验方：
[Quidway]interface serial 0   ;路由器 2
[Quidway-serial0]ppp chap user q2   ;自己路由器名
[Quidway-serial0]local-user q1 password {simple|cipher} hello   ;对方路由器名

帧中继 frame-relay：

[q1]fr switching
[q1]int s1
[q1-Serial1]ip address 192.168.34.51 255.255.255.0
[q1-Serial1]link-protocol fr   ;封装帧中继协议
[q1-Serial1]fr interface-type dce
[q1-Serial1]fr dlci 100
[q1-Serial1]fr inarp
[q1-Serial1]fr map ip 192.168.34.52 dlci 100
[q2]int s1
[q2-Serial1]ip address 192.168.34.52 255.255.255.0
[q2-Serial1]link-protocol fr
[q2-Serial1]fr interface-type dte
[q2-Serial1]fr dlci 100
[q2-Serial1]fr inarp
[q2-Serial1]fr map ip 192.168.34.51 dlci 100

帧中继监测：

[q1]display fr lmi-info[]interface type number]
[q1]display fr map
[q1]display fr pvc-info[serial interface-number][dlci dlci-number]
[q1]display fr dlci-switch
[q1]display fr interface
[q1]reset fr inarp-info
[q1]debugging fr all[interface type number]
[q1]debugging fr arp[interface type number]
[q1]debugging fr event[interface type number]
[q1]debugging fr lmi[interface type number]

启动 ftp 服务：

[Quidway]local-user ftp password {simple|cipher} aaa service-type ftp
[Quidway]ftp server enable

附录 B 实训报告模板

课程名称				编　号	
实验名称				实验日期	
院　系		专　业		班　级	
姓　名		学　号			

一、实验目的

二、实验设备

三、实验内容

四、实验过程及步骤

五、实验总结

六、教师评价